Industry Genius
Inventions and People Protecting the Climate and Fragile Ozone Layer

Stephen O. Andersen and Durwood Zaelke

Durwood Zaelke (left) is the founder and past President (1989– 2003) of the Center for International Environmental Law in Washington, DC, and Geneva. He currently is the Managing Partner in the Washington office of Zelle, Hofmann, Voelbel, Mason & Getty, as well as the founder and Director of the International Environmental Law Program at American University's law school, and the co-founder and Co-Director of the Program on Governance for Sustainable Development at the Bren School of Environmental Sciences and Management, University of California, Santa Barbara. He is the author of *International Environmental Law and Policy* (with Hunter and Salzman; Foundation Press, 2002), which has been used in more than 110 universities around the world. Mr. Zaelke is a graduate of UCLA and Duke Law School.

Stephen O. Andersen (right) is Director of Strategic Climate Projects in the US EPA Climate Protection Partnerships Division where he specializes in industry partnerships, international cooperation, and environmental performance incentives. Previously, he was Deputy Director for Stratospheric Ozone Protection. Prior to joining EPA, Dr. Andersen was professor of environmental economics at College of the Atlantic and University of Hawaii and has also worked for consumer, environmental, and environmental law non-governmental organizations. He is the author of *Protecting the Ozone Layer: The United Nations History* (with Sarma; Earthscan Publications, 2002). Dr. Andersen has a PhD in Agricultural and Natural Resources Economics from the University of California, Berkeley.

The editorial team on *Industry Genius* were:

Stephen DeCanio (University of California)
Yuichi Fujimoto (representing Japan Industry Associations)
Margaret Kerr (Consultant)
Alan Miller (Global Environment Facility)
Tetsuo Nishide (Japan Ministry of Economy, Trade, and Industry)
Paul Tebo (DuPont)
Michael Totten (Conservation International)
Dick Truly (National Renewable Energy Laboratory)

INDUSTRY GENIUS

**Inventions and People Protecting the Climate
and Fragile Ozone Layer**

**Stephen O. Andersen and
Durwood Zaelke**

Greenleaf
PUBLISHING

2 0 0 3

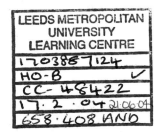
Published by Greenleaf Publishing Limited
Aizlewood's Mill
Nursery Street
Sheffield S3 8GG
UK

Printed and bound, using acid-free paper from managed forests, by
Bookcraft, Midsomer Norton, UK.

British Library Cataloguing in Publication Data:
  A catalogue record for this book is available from the British Library.
✓ISBN 1874719683

Contents

Acknowledgements

Authors Stephen O. Andersen and Durwood Zaelke walked among the giants in assembling this book. Editor Hal Kane, with the assistance of Dick Reed and Roxanne Sher-Skelton, is an absolute word-master and able to unscramble even the most vague of notions. Astronaut Richard Truly and United Nations (UN) executive Jacqueline Aloisi de Larderel put it all into perspective with their introductions. World Resources President Jonathan Lash and Global Environment Facility executive Alan Miller wrote flattering and provocative jacket summaries.

We are also indebted to our talented writing team who made particularly important contributions—Caley Johnson and Sally Rand from the US Environmental Protection Agency (EPA), Yuichi Fujimoto from the Japan Industrial Conference for Ozone and Climate Protection and Helen Tope from the Environmental Protection Authority of Victoria, Australia—who joined our travels to company headquarters, tracked down facts, and helped us see the best of the genius process.

We benefited substantially from the review by experts, but errors or omissions are entirely our responsibility. Thanks to the wisdom of our peer review team who contributed outside perspective and technical insights: Stephen DeCanio (University of California), Margaret Kerr (Consultant), Alan Miller (Global Environment Facility), Tetsuo Nishide (Japan Ministry of Economy, Trade and Industry), Paul Tebo (DuPont), Michael Totten (Conservation International) and Dick Truly (NREL). Dick Truly went beyond the call of duty by hosting a technical review in Golden, Colorado, with his extraordinary NREL engineers and managers.

The individuals who were interviewed and who made technical presentations are credited at the start of each chapter. We are particularly grateful to those in each company who had principal responsibility for setting the agenda, selecting experts, and organizing the technical presentations and factory tours. Thanks to Jay Baker and Sheila Lowe-Burke (Visteon), Giuliano Boccaletti and Fabio Borri (ST Microelectronics), Joe Clark and Dick Friel (Aviation Partners), Shuko Hashimoto and Yoshihiro Ohno (Epson), Kathleen Hogan and Maria Tikoff Vargas (Energy Star), Kenneth Martchek and Greg Craft (Alcoa), Tsutomu Okuno and Kazushige Toshimitsu (Honda), Eugene Smithart and Mike Thompson (Trane), Masaaki Yamabe and Akira Sekiya (Japan's F-Center), and Roland Caesar and Jürgen Wertenbach (DaimlerChrysler).

We thank the Center for International Environmental Law and the US EPA for making this project possible. John Stuart, Dean Bargh, and their remarkable team at Greenleaf Publishing used great skill in turning our dream into a polished book.

Preface
The genius for the next generation

Richard H. Truly

Time and again during the past century, engineering genius turned imagination into reality. We created one marvel after another: from automobiles to powered flight, from the construction of cities on Earth to the exploration of our moon and planets, from the building of our interstate highways to the creation of the Internet. But somehow along the way we neglected and even jeopardized the prosperity that flows from natural ecosystems by creating a gargantuan demand for energy for our rapidly growing human population.

In April and May of 1961, Russian Yuri Gagarin and American Alan Shepard became the first humans to see the whole Earth and to feel the overwhelming realization of its fragility. The first astronauts saw the forces of nature and climate—smoke from volcanoes, spiraling hurricane clouds, and the blowing dust of the Sahara—and also saw human wonders—the Great Wall of China, the pyramids, and city lights.

Just 20 years later in November, 1981, on my first space mission aboard *Columbia*, I marveled at similar wonders—but I also saw the sobering evidence of thick pollution, from the Los Angeles basin to Mexico City to Tokyo. Twenty years after my first flight, US astronaut Frank Culbertson, Commander of Expedition Three aboard *Space Station Alpha*, reported that Earth was a planet clouded in smoke and dust and visibly scarred by mining and forest destruction. Culbertson believes the changes are a cause for great concern, and I couldn't agree more.

Our high level of energy consumption today reminds me of Edna St. Vincent Millay's famous stanza:

> My candle burns at both ends
> It will not last the night
> But ah, my foes, and oh, my friends—
> It gives a lovely light!

> *(A Few Figs From a Few Thistles*, "First Fig")

It is, indeed, a lovely light. The result, however, is the continuing depletion of our storehouse of fossil fuels, our increasing reliance on vulnerable infrastructures, and our continuous addition of waste to our environment.

I believe in a transition strategy called "investing for the future." This strategy is the quick and substantial improvement of the efficiency of both energy production and end use, as well as mitigation of the environmental consequences of fossil fuel use and of nuclear waste. This strategy relies on the use and further development of renewable sources of energy. "Investing for the future" requires preparation and work on our part to move from incremental changes in the near term to full paradigm shifts in the long term. Paradigm shifts require unprecedented engineering, public involvement, and infrastructure.

Imagine a world described as YIMBY—"Yes In My Back Yard"—welcoming windmills, solar collectors, and complete product recycling centers. Imagine communities and businesses joining together to build the infrastructure necessary for the use of clean hydrogen power. Worldwide, the pay-off is that people will be able to prosper while putting minimal burdens on future generations, and that all countries will become more self-sufficient and less likely to wage war. No small feat!

This book is full of examples of how this imagined future is already becoming reality. It reminds us that human genius and corporate leadership can help citizens craft a future for their children and grandchildren. It reminds us that environmental progress also depends on research, government policy, and market demand. The genius in this book is a beacon for the next generation of engineers who will solve the daunting challenge of climate change with energy efficiency and renewable resources.

If we stay this course, then public policy, science, and engineering will transform markets into users of clean, secure, reliable energy. The candle will last the night!

Vice Admiral Richard Truly (Ret.) is Director of the Department of Energy's National Renewable Energy Laboratory (NREL) and Executive Vice President of the Midwest Research Institute. NREL is a center of excellence for renewable energy and energy efficiency research to benefit both the environment and the economy. Prior to joining NREL, Truly was Vice President of the Georgia Institute of Technology, and he served as the National Aeronautics and Space Administration (NASA)'s eighth administrator under President George H.W. Bush from 1989 until 1992. In 1986 he led the investigation of the *Challenger* accident and spearheaded the painstaking rebuilding of the Space Shuttle program. Truly's career as an astronaut included work in the Manned Orbiting Laboratory and NASA's Apollo, Skylab, Apollo-Soyuz, and Space Shuttle programs. He lifted off in November 1981 as pilot aboard *Columbia,* and he commanded *Challenger* from August to September 1983. Richard Truly and his senior NREL engineers provided technical review to this book and are nominating additional technologies for subsequent editions.

Foreword
Just when Earth needed genius

Jacqueline Aloisi de Larderel

The world is changing, it has always been changing, and experience tells us that the companies that will be on the scene tomorrow are those that are adapting, not those that are resisting change. This book is about the companies that celebrate change and welcome challenge, and the geniuses in those companies who are finding their way around the laws of physics, confidently, stubbornly, and relentlessly, in pursuit of both profit and environmental protection.

For the past 15 years my workplace has been the United Nations Environment Program (UNEP), where I directed the Division of Technology, Industry, and Economics (DTIE). Our mission is to encourage decision-makers in government and industry to develop and adopt policies, practices, and technologies that are cleaner and safer, and which make efficient use of natural resources. From this perspective, it is easy to see that, if current trends in consumption of natural resources continue, we would need three or four planets by the end of the century. It is also easy to see that the best way to address these pressing problems is to mobilize the corporate sector, where both companies and society can benefit.

Fossil fuel resources are being depleted, pollution is still imperiling human health and ecosystems, and greenhouse gases are affecting the climate. Sustainable development is the strategy of moving to a global society based on abundant renewable energy, cleaner production, ecosystem approaches to commerce that emphasize the recovery and recycling of materials, and prosperity that does not put our natural capital at risk. National and global populations demand changes toward sustainability, and they are growing impatient.

Fortunately, government and industry have come a long way in the last few decades and can act quickly to build on successes already achieved by working together. Governments have been removing perverse subsidies and insisting that polluters pay the external costs of environmental despoilment. Leading private-sector companies have gradually shifted from expensive and reactive end-of-pipe compliance to a cost-effective approach of pollution prevention, cleaner production, and eco-efficiency. It is no coincidence that most of the companies featured in this book are at the forefront of environmental leadership. Such leadership enthusiastically supports the geniuses whose ideas will lead to profits from new green products.

Concern over global climate change has led to a number of international political responses, most notably the United Nations Framework Convention on Climate Change and its Kyoto Protocol. When it enters into force in early 2003, developed countries "party to the Kyoto Protocol" will be required to reduce their emissions of carbon dioxide (CO_2) to agreed targets. Governments will achieve compliance with the Kyoto Protocol by changing the operating rules of business by applying a mixture of command-and-control regulations, energy taxes and price incentives, market transformations, and voluntary programs. In addition to providing incentives for the development of renewable energy supplies and energy-efficient products, the Protocol creates new business opportunities through its provisions for emissions trading, the clean development mechanism, and joint implementation.

Let me provide some striking signs of change and examples of success in protecting the climate. Vehicle manufacturers in Europe have agreed to improve average fuel efficiency by 25% by 2008, setting a cap of just 140 grams of CO_2 emitted per kilometer traveled (8 oz per mile). Some manufacturers are now marketing tires that can reduce energy consumption, and, if these tires are used everywhere in the world, then CO_2 emissions will be reduced by about 60 million tons per year! Denmark, Germany, The Netherlands, and the United Kingdom (UK) are implementing "greenpower" policies that will drastically increase the share of renewable energy in the overall energy mix; and France is now following the same path. Pioneering regulations by Austria, Denmark, and Switzerland are encouraging the European Commission to consider stringent regulations, or even bans, on potent hydrofluorocarbon (HFC) greenhouse gases. Emerging economies also understand the need to move toward more sustainable production and consumption patterns. For example, by improving technologies and reducing wasteful practices, China is managing to significantly "decouple" economic growth from energy consumption.

The environmental and social responses to climate change will bring change to the energy industry and its customers in the 21st century. By continuing to spur technological innovation, these responses will have an increasing impact on the way that energy is generated, used, and even conceived. For forward-looking companies in all parts of the world there is an unparalleled opportunity to develop creative approaches to generating and using energy, and the winners will be those who create value without environmental costs.

This book presents a number of inspiring examples of technologies and products designed for sustainable development; and there are many others that deserve to be reported as well. We need many more innovations to move us away from current, unsustainable trends.

You, the reader of this book, can make a difference. I urge you to suspend your doubts, to open your mind, and allow yourself to dream of a better world. Unleash the genius in yourself and your colleagues. Change your own home and lifestyle and rededicate your workplace to working for the good of both present and future generations.

Jacqueline Aloisi de Larderel was Executive Director of the Division of Technology, Industry, and Economics of the United Nations Environment Programme (UNEP) from 1987 to 2003. Her Center brings together industry, government, and non-governmental organizations to work towards environmentally sound industrial development using environmental management tools and the Cleaner Production concept. Before joining UNEP, she was employed by the French Ministry of the Environment. Ms. Aloisi de Larderel holds a Master of Science degree in Chemistry and Pharmacology from the University of Paris and Master of Business Administration from the European Institute of Business Administration.

Introduction

Stephen O. Andersen
and Durwood Zaelke

"The Wright brothers flew right through the smokescreen of impossibility."

Charles Franklin Kettering

Be inspired

In the stories that follow you will be amazed by what these geniuses are accomplishing for the environment. You will be inspired and informed when you see how these inventors dismiss skepticism, abandon perceived limits, inspire teamwork, and work "outside the box." Apply their techniques to your own observation and discovery, to become more alert, more creative, and more interesting. What would you design if there were no limits? Would you have the courage to announce your goals to management, colleagues, suppliers, customers, family, and others who might share your dream?

We have all kicked ourselves for not thinking of one invention or another. Or we may have had the idea, but simply not acted on it—leaving us with a sense of regret that we failed to move forward. Consider how these stories of creative genius might help you in the future. After you read the Visteon story, disassemble a product in your mind and put it on the Visteon "yellow board." What do you see? After you read the Seiko Epson story, visualize Epson's "keystroke-powered personal computer." What limits the sustainability of your product? After you read Trane's story, imagine Trane's ceramic bearings and consider a world without friction. Where will you go? The stories in this book show that genius often springs from common sense and even from the intuitively obvious.

Watch for accidental and strategic genius

This book tells the story of eight companies and two government enterprises that are using their inventive genius to protect the climate and the ozone layer. Some results of this genius are almost accidental—product engineers questing after profit or technical elegance who stumble onto fuel efficiency. More often, though, this genius results from management and leadership recognizing that customers demand environmental protection and that profits increase with pollution prevention and care for the future. And, in a surprising number of cases, genius results from self-motivation that comes from personal concern for our Earth and its environment.

Aviation Partners attracted geniuses to pursue new ideas of physics and aeronautical engineering that incidentally reduced greenhouse gas emissions. Their stories show that smart science and engineering can often lead to environmental protection.

Meanwhile, Seiko Epson, Trane, and Visteon executives assigned the geniuses who had spearheaded the successful elimination of ozone-depleting substances to tackle the challenge of climate protection. The executives did this because the engineers' success with ozone-safe technology had increased profits, improved product performance and reliability, and built company reputation—all simultaneously.

ST Microelectronics recruited genius when its CEO's environmentalist son challenged the way that business had ignored the environment. The CEO and his idealistic son persuaded the company to create comprehensive environmental policies and to involve all employees in its quest for sustainability.

Alcoa organized its geniuses after the US EPA approached aluminum manufacturers with an offer of a global partnership to investigate methods for reducing the emissions of powerful perfluorocarbon (PFC) greenhouse gases. Alcoa's geniuses promptly identified unexpected profits from reducing PFC greenhouse gas emissions—at a time when environmental protection was becoming a higher corporate priority.

The legendary entrepreneurs who founded Honda and Seiko Epson had an abiding respect for nature. These companies profit from avoiding manufacturing waste, from brand reputation built on environmental leadership, and from products marketed successfully on the advantage to customers of energy efficiency. Future engineering geniuses with environmental ambitions know that these are two of the best companies that they can work for.

Unleash your creativity

The story of the hero overcoming challenges on the way to an important goal is an archetype as old as language itself. Such stories continue to teach and inspire today. We hope that the stories of the genius-masters in this book will earn a place in this canon, alongside other important recent books. These works include Michael J. Gelb's *How to Think Like Leonardo da Vinci*,[1] a step-by-step training in imagination, visualization, description, drawing, and perfecting, and Annette Moser-Wellman's *The Five Faces of Genius*,[2] which classifies the creative techniques of past and present geniuses as seer, observer, alchemist, fool, and sage and presents exercises to strengthen these talents in all of us. Our goal is to acknowledge the geniuses working for the protection of the Earth and to tell their stories, so that you might unleash your own creativity.

Be a genius talent scout

Creating a sustainable future is a big job and we need your help. Be a talent scout and recommend environmental genius that will be featured in the next edition. Your genius nominations will be considered with three companies that earned honorable mention in this volume: AeroVironment, Michelin, and Turbocor.

AeroVironment has demonstrated sun-powered flight with ascent to altitudes of over 95,000 ft (30,000 m), shattering the altitude records of both propeller and jet-engine aircraft. AeroVironment's founder, Paul MacCready, is the aviation genius who developed the *Gossamer Condor* and the *Gossamer Albatross*, which made the first person-powered flight across the English Channel.[3]

Michelin is the first company to provide no-compromise safety and environmental tire performance. Michelin Energy tires are a "top runner" in fuel economy, among the best in high traction, and they achieve low noise. Some 60 million tones of CO_2 emissions could be reduced each year if all tires matched this performance.[4]

Turbocor developed a compact, oil-free air-conditioning compressor with magnetic bearings that is up to 30% more efficient than the competition—smaller, lighter, and quieter.[5]

It's the environment, stupid

The old thinking was that a healthy environment was a luxury good that only the rich could afford. The new thinking is that the natural environment is a fragile powerhouse of wealth and satisfaction, that environmental protection is prof-

---

1   Michael J. Gelb, *How to Think Like Leonardo da Vinci* (New York: Delacort Press, 1998).
2   Annette Moser-Wellman, *The Five Faces of Genius* (London: Penguin Books, 2001).
3   See www.AeroVironment.com.
4   See www.Michelin.com.
5   See www.Turbocor.com.

itable, and that government–industry partnerships can help transform markets to guide producers and consumers to the right balance.

We now know that our prosperity depends as much on the outputs of natural ecosystems as it does on human and manufactured capital. This critical lesson is presented persuasively by Paul Hawken, Amory Lovins, and Hunter Lovins in *Natural Capitalism*.[6] We can now visualize production where no resource is wasted (100% product and zero waste) and where all materials are recycled continuously, as William McDonough and Michael Braungart describe in *Cradle to Cradle*.[7]

We have also learned that many of humankind's best inventions were inspired or copied from nature, and that scientific advances allow us to look deeper and deeper into nature to unlock new products that bypass poisons, pesticides, and hazardous waste. This is described vividly in Janine Benyus's *Biomimicry: Innovations Inspired by Nature*.[8]

Smart companies are poised to profit as consumers increase their demand for green products, and as citizens increase their demand for environmental quality. Gretchen C. Daily and Katherine Ellison describe this new strategy for profitability in *The New Economy of Nature: The Quest to Make Conservation Profitable*.[9] And, in *Walking the Talk: The Business Case for Sustainable Development*,[10] Chad Holliday, Stephan Schmidheiny and Philip Watts, who head DuPont, Anova Holding, and Royal Dutch Shell respectively, pronounce that future economic growth will be promoted by solving environmental and social problems. The bottom line is that what's good for the environment is good for business.

Protecting the Ozone Layer: The United Nations History by Stephen O. Andersen and K. Madhava Sarma[11] documents how a global partnership—between governments, business, and civil society—was able to protect the stratospheric ozone layer before it was too late. And *International Environmental Law and Policy* by David Hunter, James Salzman, and Durwood Zaelke[12] documents the global quest for environmental protection and sets forth the legal liability for jeopardizing sustainable development. These books are all remarkable resources that we can use on our way to creating our own industry genius.

6   Paul Hawken, Amory Lovins, and L. Hunter Lovins, *Natural Capitalism: Creating the Next Industrial Revolution* (Boston, MA: Little Brown, 1999).

7   William McDonough and Michael Braungart, *Cradle to Cradle: Remaking the Way We Make Things* (San Francisco: North Point Press, 2002).

8   Janine Benyus, *Biomimicry: Innovation Inspired by Nature* (New York: William Morrow, 1997).

9   Gretchen C. Daily and Katherine Ellison, *The New Economy of Nature: The Quest to Make Conservation Profitable* (Washington, DC: Shearwater Books, 2002).

10   Charles O. Holliday, Jr., Stephan Schmidheiny, and Philip Watts, *Walking the Talk: The Business Case for Sustainable Development* (Sheffield, UK: Greenleaf Publishing, 2002).

11   Stephen O. Andersen and K. Madhava Sarma, *Protecting the Ozone Layer: The United Nations History* (London: Earthscan Publications, 2002).

12   David Hunter, James Salzman, and Durwood Zaelke, *International Environmental Law and Policy* (New York: Foundation Press, 2nd edn., 2002).

1

Alcoa Aluminum
Putting energy in the bank[*]

Innovations in the manufacture of aluminum are transforming it from a major source of greenhouse gases into a solution for increasing energy efficiency, recapturing energy and materials, and reducing pollution—with big financial and environmental benefits as the world realizes that aluminum is a climate protection metal.

The next time you drink a Coca-Cola and recycle the can, you can take heart that you are saving 95% of the energy needed to make a new can from aluminum bauxite. This is because once you make aluminum it becomes, in effect, an energy bank that you can tap into over and over again.

When we recycle aluminum, we create the potential to save 20 billion kilowatt-hours of electricity. That is *one per cent* of the electric power used annually in the USA That should make you want to treat that empty can with some respect.

The new environmental reality is that re-use of natural resources takes us one step closer to sustainability—saving the planet from resource depletion. "Waste not, want not."

One of the most interesting and untold stories illustrating aluminum's global importance occurred during the early days of World War II. Late one night in 1942, a German U-boat off the coast of Long Island surfaced and a group of four German saboteurs disembarked, intent on changing the course of the war by destroying America's aluminum smelters in New York and Tennessee. Their plan was to disrupt the electricity supply long enough for the molten metal in the furnaces to solidify, thereby closing the smelters down. Only blasting could unfreeze the furnaces at that point. And the effects of the blasting would take months to repair, causing irreparable damage to the vital war effort.

* The authors are grateful for the interviews and supplementary assistance of Patrick Atkins, Kerry J. Farmer, Cheryl Kirkland, Greg C. Kraft, Kenneth J. Martchek, Steven H. Myers, and Alton T. Tabereaux (Alcoa USA); Ken Mansfield and Jim Harpley (Alcoa Portland Australia); and Krista Milne and Helen Tope (EPA Victoria Australia).

Luckily, the Federal Bureau of Investigation (FBI) uncovered the plot and caught the saboteurs before they could strike. This ubiquitous metal, previously known mostly in the kitchens of America, was being transformed into lightweight vehicles and aircraft that could travel farther on each gallon of fuel, and which were less vulnerable to fire. At that time, aluminum had only been manufactured commercially for about 75 years, and the world was just beginning to realize how important this abundant resource would become in every aspect of our daily lives.

The new metal on the block

Aluminum, or "aluminium" as it is spelled in most countries, is one of the most versatile products in the world. And it is one of the "youngest" metals in terms of commercial development. It can be rolled to a thickness of five microns for delicate applications, and yet it can be used as armor-plating or to add strength and stability to skyscrapers. It is produced in nearly every region of the world, made from one of the Earth's most abundant minerals, and can be recycled infinitely. It retains its performance properties at very low temperatures and can be sprayed on polymers to make lightweight, heat-retaining blankets for use in emergencies. Reduce it to powder and it becomes rocket fuel.

Aluminum is also making a key contribution to climate protection because it enables lower-weight, fuel-efficient engines and parts to be built for cars and trucks as well as for high-speed rail and sea travel. Aluminum is corrosion-resistant and low-maintenance. It is used to transmit and distribute electricity over long distances. Aluminum packaging is integral to food preservation. And, as stated above, a key environmental feature is that aluminum products serve as an energy "bank," since recycling requires only 5% of the energy needed for primary metal production.

Unlike gold, silver, iron, copper, and bronze, which have been used since the dawn of civilization, aluminum is a young metal. Aluminum has only been produced on an industrial scale since 1886, when American Charles Martin Hall and Frenchman Paul L.T. Héroult independently and virtually simultaneously developed a process of using electric current to reduce alumina to aluminum in a molten solution known as cryolite. Today virtually all of the world's aluminum is produced via the

THE ADVANTAGES OF STRENGTHENING VEHICLES WITH LIGHTWEIGHT ALUMINUM

- Aluminum has only one-third the density of steel. When aluminum replaces steel in automotive applications, only 1 lb (0.45 kg) is typically needed to perform the same function of 2 lb (0.9 kg) of steel. The aluminum vehicle is lighter and stronger.

- Ancillary components, such as suspension parts, can be smaller and lighter because they carry less load. The Audi all-aluminum A8 sedan realized primary and secondary weight-savings of 1,100 lb (498 kg) compared to a similar steel vehicle, for example.

- The use of aluminum allows multiple parts to be cast into one assembly, saving weight and reducing the cost of assembly, fasteners and welding.

Figure 1.1 Typical aluminum "pots" in smelter hall

Hall–Héroult electrolysis process. Alcoa was founded as the Pittsburgh Reduction Company in 1888 by Charles Martin Hall in Pittsburgh, Pennsylvania. Its name was changed to the Aluminum Company of America in 1906, and changed again to Alcoa, Inc., in 2000.

This modern miracle had a price

Producing this miracle metal has environmental consequences. The commercial process currently used worldwide requires large amounts of energy and produces various pollutants, including sulfur oxides (SO_x) and perfluorocarbons (PFCs), which are among the most potent greenhouse gases emitted by human activity.

The good news is that companies are motivated to reduce energy consumption, which now accounts for 25% of primary aluminum production costs. And the industry is striving to be a pioneer in sustainable development, as well as part of the climate solution. Alcoa is leading the charge by reducing direct and indirect emissions of greenhouse gases from the century-old Hall–Héroult production process (see Figure 1.2), and is working to revolutionize the industry by commercializing the world's first carbonless "inert anode." If successful, the inert anode— the aluminum industry's Holy Grail—will eliminate SO_x, PFCs, and other carbon-based emissions.

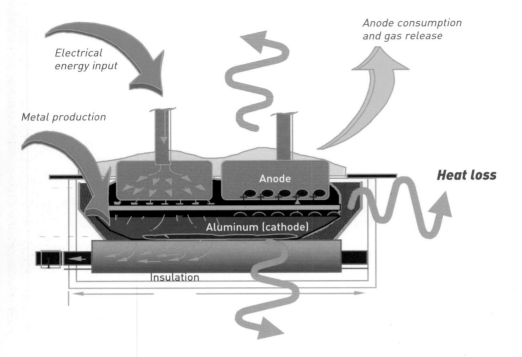

Anode consumption
and gas release

Electrical
energy input

Metal production

Anode

Heat loss

Aluminum (cathode)

Insulation

Figure 1.2 Typical Hall–Héroult production process

> Both community and governments are challenging us to go beyond our
> traditional role of efficiently meeting their economic needs—they are
> telling us that it must be done with a smaller environmental and social
> footprint (John Pizzey, Executive Vice President of Alcoa).[1]

Alcoa has also established an aggressive long-term environmental program
based on sustainable development principles, and has set ambitious goals to guide
the company through the year 2020. Alcoa uses sustainable development as a
concept to address how it does business, how it interacts with its plant communities and other stakeholders, and how it looks at public policy issues.

For example, Alcoa actively promotes the advancement of public policy for
climate protection, recently stating in a 2002 letter co-authored by the National
Wildlife Federation to the US Congress: "the risk of significant climate change is an
issue of vital importance requiring action." For Alcoa, addressing climate not only
makes long-term economic sense; it is also necessary for the company to act now
in order to ensure that it is well positioned for future growth.

1   Annual Meeting of the Aluminum Association, Nemacolin, PA, September 30, 2002.

Aluminum was once part of the problem

Electrical energy is critical to the smelting process, and its use has always been a major cost factor to the aluminum industry. Its cost has been so large that primary production sites are typically located where energy is the least costly, rather than where alumina-rich bauxite is the most abundant. The majority of the world's aluminum production is located near hydroelectric resources, which help ensure consistent supplies of low-cost electricity.

Steady progress has been made over the last century to reduce energy consumption from over 20 kWh of electricity to produce a pound (0.45 kg) of metal at the turn of the century to less than 6 kWh per pound in today's most modern plants. However, the industry is still high on the list of the most "energy-intensive" manufacturing processes.

Today, Alcoa is continuing to reduce energy use by process optimization, by reducing energy use in products through the use of improved alloys, by striving for higher aluminum recycling rates, and by innovative forming and product design to provide the same performance with less aluminum. For example, a 12 ounce beverage can made of aluminum contains 30% less aluminum today than it did just 20 years ago, but provides the same food protection and convenience to the customer. In October, 2002, Alcoa announced a new corporate sustainability goal that 50% of its products will be made from recycled aluminum by 2020, except for raw ingot that is sold directly to others. Figure 1.3 shows a comparison of the CO_2 savings of recycled versus new aluminum.

The aluminum industry has always pursued technology to reduce the energy intensity of its operations. The discovery of PFCs, however, came as a recent and unpleasant revelation in the 1980s, when scientists studying climate change identified these chemicals as highly potent and persistent greenhouse gases. Unlike the more abundant greenhouse gases in the atmosphere—carbon dioxide (CO_2), nitrous oxide (N_2O), and methane (CH_4)—that have both natural and man-made sources, PFC emissions appear to come mainly from industrial activities. Unfortunately for the aluminum industry, the manufacture of the metal turned out to be a primary source of PFCs, as summarized in a report by R.A. Rasmussen (1983).[2]

GREENHOUSE GAS POTENCY

Greenhouse gases are not all equal in how they affect the climate. Each greenhouse gas differs in its ability to trap heat in the atmosphere and how long it stays in the atmosphere. Any useful inventory of greenhouse gas emissions must include all of the chemicals and in some way reflect the potency and persistence of the various gases. Because CO_2 is the major human cause of global warming, it is used as the reference for comparing the potency of other greenhouse gases. The "global-warming potential" (GWP) of CO_2 is set at "1." GWP is a unit of measurement of the atmospheric warming that will occur relative to the warming caused by the same quantity of carbon dioxide. It is always expressed for a specific time-period, usually 100 years, even though some gases last longer than that in the atmosphere and others have a shorter lifetime.

---

2   R.A. Rasmussen and J.E. Lovelock "The Atmospheric Lifetime Experiment. 2. Calibration," *Journal of Geophysical Research* 88.C13 (1983): 8,369-78.

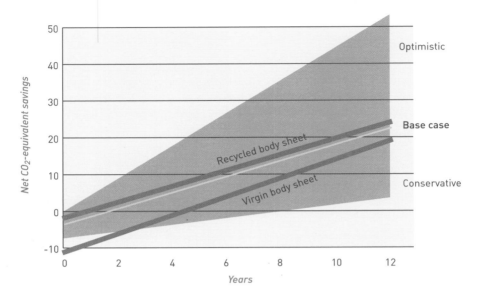

Figure 1.3 CO_2 saving of recycled versus newly manufactured aluminum

The PFCs emitted inadvertently from the primary aluminum production process—CF_4 (carbon tetrafluoride [tetrafluoromethane]) and C_2F_6 (hexafluoroethene)—
have 100-year GWPs estimated to be 6,500 and 9,200 (see box on previous page)
and atmospheric lifetimes of at least 50,000 and 10,000 years respectively. The
GWPs for PFCs get much higher relative to CO_2 as the time-horizon for comparison
is extended. This is because longer time-periods account for more of the impact: for
example, while the 100-year GWP for CF_4 is 6,500, its 500-year GWP would be
10,000.

Most countries that are developing climate-change strategies are using a multi-
gas or "basket" approach to reduce emissions of the six principle greenhouse gases,
including PFCs.[3] Policies based on the basket of gases are more cost-effective than
reducing just one or two gases, and can better reflect national circumstances.
Basket or no basket, it is obviously a good climate strategy to address the most
persistent gases as soon as possible to avoid an unnecessary build-up of chemicals
that will remain in the atmosphere for 10,000–50,000 years. Currently, PFCs
account for an extremely small portion of greenhouse gases in the atmosphere (81
parts per trillion [ppt] for CF_4 and 3 ppt for C_2F_6 compared with 360 parts per
million for CO_2), and emissions are well below 1% of total annual global emissions
on a CO_2-equivalent basis. But early action on reducing PFC emissions will result in
important long-term benefits.

---

3   The six gases are CO_2, PFCs, sulfur hexafluoride (SF_6), hydrofluorocarbons (HFCs),
methane, and N_2O.

How the US EPA became a partner with Alcoa

Predictably, the US EPA (Environmental Protection Agency) began contacting the primary aluminum producers in the USA, including Alcoa and Reynolds Metals (now part of Alcoa), shortly after the release of the first scientific report linking PFC emissions to aluminum production. The people at Alcoa were concerned. It wasn't that Alcoa didn't want to be a good corporate citizen and help protect the environment, but it didn't have a lot of confidence that any regulator would ever understand its business or be a constructive partner in solving the problem. And, without understanding the aluminum production process—which is still in many ways as much an art as it is a science and technology—the EPA would surely complicate or distract any attempt to improve the process.

But the EPA's communication was not the usual message from the office of enforcement threatening legal action. It came instead from the office that deals with voluntary industry partnerships. Cindy Jacobs, the EPA's first program manager for the Voluntary Aluminum Industry Partnership (VAIP), recalls what happened. "I had just read the Rasmussen report on the discovery that PFCs were emitted from aluminum production, and I knew that this was a problem that EPA had better address immediately."[4] PFCs were targeted as a priority due to their long atmospheric lifetimes. Jacobs says:

> **ALUMINUM PRODUCTION DETAILS**
>
> An electric current is passed through a molten solution of alumina and cryolite (sodium aluminum fluoride) in a series of reduction cells (also called "pots") that are lined at the bottom with carbon (the cathode). Inserted into the top of each pot are carbon anodes, the bottoms of which are immersed in the molten solution. The electric current causes the oxygen from the alumina to combine with the carbon of the anode. The molten aluminum settles at the bottom of the pot and is drawn off by tapping into a vacuum crucible. The molten aluminum is then transferred to furnaces for alloying and casting into various shapes.

> My EPA colleagues had pioneered voluntary partnerships for ozone-layer protection and for energy efficiency through the Green Lights and Energy Star programs. They train us how to listen, how to problem-solve, and, most importantly, how to become an asset to companies seeking change for the better.[5]

Dr. Alton Tabereaux, Alcoa Process Technology Manager, who at the time was at Reynolds Metals, said "EPA scared us to death." "We were not looking forward to the first meeting," he added. "Imagine our surprise when Cindy offered to help us pursue whatever was needed to reduce PFC emissions." "Together, we developed a basic research program to determine how and why the PFC-generating phenomenon known as 'anode effects' got triggered and how to measure PFC emissions." "EPA won us over," Dr. Tabereaux admitted. "Before long we were encouraging other companies to participate and share information on successful techniques to reduce emissions." The Aluminum Association provided a forum for US companies

---

4   Personal interview with Cindy Jacobs, US EPA, currently Chief of the Market Sectors Section of the Energy Star Commercial and Industrial Branch.
5   *Ibid.*

to work cooperatively with the EPA on PFC emissions reductions through the VAIP. Similar efforts were initiated by other aluminum associations throughout the world as companies stepped up to "globalize" the effort.

PFC studies reveal unnecessary waste and lost profits

"Before Rasmussen's and other scientific publications, our chemists told us not to worry about PFCs because they are not toxic," remembers one of Alcoa's managers. "Then we found out that they are potent greenhouse gases." PFCs are emitted only occasionally, when the alumina content of the molten cryolytic bath containing fluoride falls below a critical level. Below this critical level, the pot voltage rises rapidly and the fluoride in the bath is transformed through electrolysis into PFCs (see Figure 1.4).

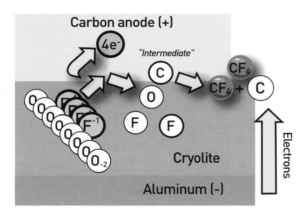

Figure 1.4 Formation of PFCs during anode effect

Source: M. Dorreen, D. Chin, J. Lee, M. Hyland, and B. Welch, *Light Metals* (Orlando, FL: Elsevier Science, 1998)

PFCs escape the pot hooding, into the exhaust system, through the scrubbers and into the atmosphere. Anode effects are not a continuous part of the process, but rather a periodic and often random phenomenon in aluminum reduction cells. The frequency and duration of the effects depends largely on the type of technology used and on work practices employed for the cells. A pot enters anode effect status when it exceeds 8.0 volts. It is considered extinguished when the voltage falls below 6.0.

Although the industry did not know the amount of PFCs that were emitted from the process, they already knew something about the anode effects that generated them. Occasional anode effects were at one time considered useful by the cell operators, because they established a known concentration of alumina in the bath to reset the feed control to avoid excessive feeding and the formation of alumina sludge in the pot. The anode effect and actions to stop it also create turbulence that was thought to help dislodge alumina sludge in the bottom of the pot, allowing that sludge to be reduced to aluminum, as well as removing excess carbon dust from under the anode electrodes.

Although aluminum production has used essentially the same materials and process for the past century, it is still somewhat mysterious, and careful tending is required to operate the process in an efficient manner. The process requires a regular replenishment of alumina to be fed into the pot, the occasional addition of fluoride salts, along with regular drawing down or "tapping" of the molten aluminum that has been produced into a crucible. The typical procedure to replenish alumina has been to dump in large quantities at scheduled intervals.

The investigations into PFC emissions determined that alumina concentration in many pots became depleted before the next scheduled batch feed operation, resulting in the electrolysis of fluorides and production of PFC gases. The PFC gases formed an electrically resistant gas film under the anode electrodes, forcing the voltage up from the desired 4–5 volt range to over 30 volts and driving the process into anode effect. When this happened, the pot liquid materials would become very hot and turbulent, which does not maximize efficiency of aluminum production.

The work with EPA confirmed that, when a pot is on anode effect, raw materials are consumed but aluminum is not produced. Alcoa's original method to detect when a pot had gone into anode effect was to watch for light bulbs attached to the pots to turn on when the voltage spiked. A floor worker would literally run to the pot when the light bulb came on and thrust a long pole cut from green wood into the 1,800°F (982°C) liquid mixture and add additional alumina powder. The burning moist pole would create bubbles that caused the aluminum metal to short circuit to the anodes, which would end the anode effect and help restore the balance in the pot.

Today the industry relies increasingly on sophisticated computer controls to detect and extinguish anode effects quickly and to document the time and chemistry required to end the anode effect. Across the industry, process techniques have been adopted to monitor and control the concentrations of alumina in the bath more effectively, such as shifting from batch-feed to computer-operated, point-feed technology that allows feeding alumina to individual pots at different rates as needed.

A legend in his own time

Dr. Tabereaux is the aluminum scientist who first calculated the PFC emissions from aluminum production based on scientific principles and before plant measurements. The "Tabereaux Equation," as it is commonly referred to by the industry, was published in 1994 in the *Journal of Metals*. Tabereaux is a paragon in

Alton Tabereaux

the aluminum business; whatever he says is gospel. His current job is to be the "super process technology manager" of Alcoa's worldwide facilities. This super management job is a testament to the high regard that Alcoa has for Tabereaux.

Tabereaux doesn't do his job from a desk. He explains, "managing aluminum production requires you to be on-site, talking to the plant management and staff. You have to look at the pots, and see how the chemical bath is being managed." This means that he spends almost all of his working days on the road, although he is quick to say that the company is very good about encouraging him to get home for the weekend to be with his family in Alabama. "I've worked for other companies that insisted you stay over Saturday night to get a lower fare," he recalls.

Tabereaux is a soft-spoken guy, about 5 ft 9 in, with short brown hair, probably in his mid-fifties. He was a Marine recruit doing his boot camp days at Parris Island, South Carolina, in 1970, and went on to earn his PhD at the University of Alabama in 1981 in metallurgy. In the late 1980s and early 1990s he ran the Extractive Metallurgy Department in the manufacturing technology research lab for Reynolds Aluminum Company. He has earned 15 US patents regarding smelter technology, authored over 50 technical publications, and is a frequent lecturer at international aluminum technology courses in Norway, Australia, Canada, and the USA.

He received the prestigious *Journal of Metals* Award from the Minerals, Metals, and Materials Society in recognition of the most notable technical paper published both in 1994 and 2002. He became an Alcoa employee when Alcoa purchased Reynolds in 2001, but has been involved in the EPA VAIP since its launch in 1995.

AWARD-WINNING ALCOA PORTLAND PROCESS RECYCLES WASTE INTO ALUMINUM*

Alcoa-operated Portland Aluminium** in Victoria, Australia, has developed the "Alcoa Portland SPL Process" to convert hazardous spent pot lining (SPL) waste into valuable materials for re-use. With this process, fluoride is recycled as a raw material in smelting aluminum, cyanide is destroyed, carbon is used to generate process heat, and the remaining granulated slag is approved by the Environment Protection Authority Victoria (EPA Victoria)[†] for unrestricted environmental uses such as road-making and use in concrete aggregates.

The pot liner used in the aluminum-smelting process is composed of carbon and refractory brick held in a steel pot, which contains the molten aluminum and the electrolyte (sodium aluminum fluoride). A pot is normally used for three to eight years until the inside liner cracks or erodes and the pot fails. The remaining "spent" carbon and refractory lining is hazardous waste—containing fluoride and cyanide.

In most countries, SPL is considered hazardous waste and is increasingly being banned from landfill disposal. SPL has been stockpiled because there has been no acceptable technical and economic treatment process. The Portland Aluminium smelter generates about 5,500 metric tons a year of SPL, and Alcoa's Point Henry smelter generates another 3,500 metric tons. The North American aluminum industry generates more than 100,000 metric tons a year; and global output is around 500,000 metric tons.

Ken Mansfield, Portland Aluminium's Manager of the SPL project, said, "The development of the Alcoa Portland SPL process was initiated when Portland Aluminium took an unconventional decision in the aluminum industry in the mid-eighties and abandoned landfill disposal in favor of finding a more productive solution for SPL waste." In 1989, the then Portland Aluminium Plant Manager, David Judd, had a vision to make Portland Aluminium's smelting environmentally invisible and sustainable—a "smelter in the park"—and he sent a team of experts across the globe to find the best processes for treating SPL.

The team studied processes and brainstormed with experts at Comalco, Lurgi, KHD, Elkem, and Reynolds. But none of the existing processes satisfied Judd's demanding environmental and economic criteria. "David had the support of Alcoa and the other owners of Portland Aluminium to make this a top priority, and a team of our own best experts collaborating with experts from other organizations and operating with an open-ended budget," said Ken Mansfield.

Success came in 2001, after ten years of daunting research, development, and problem-solving in collaboration with Ausmelt[††] and the Commonwealth Scientific and Industrial Research Organization (CSIRO).[‡] "Unfortunately, David Judd died before his vision was realized. However, he would have been proud of all who contributed to the Alcoa Portland SPL process and that it achieved his demanding criteria for treating the hazardous SPL," continued Mansfield.

EPA Victoria Chairman Mick Bourke said, "We've been working with Portland Aluminium in partnership with the local community for over six years, and this is a great outcome for all concerned. The Alcoa Portland SPL process is a world-leading example of sustainable development and the innovation of Victorian industry."[‡‡]

In 2002, Portland Aluminium was awarded the prestigious Banksia Foundation Environmental Award and the Premier's Business Sustainability Award for Victoria in recognition of its groundbreaking innovation. When the Premier of Victoria, Steve Bracks, announced the award, he stated that "Government's role is to hold a beacon to a sustainable future and to create the conditions for business to respond to the opportunities this future presents."

* This case study was developed by Ken Mansfield, John Lippelgoes, Gillian MacMillan, and Joan McGovern from Alcoa Portland (Victoria, Australia) and by Helen Tope, with the assistance of Krista Milne, Scott Maloney, Lyn Denison, Dirk Dukker, Kate Noble, and Tony Robinson from the Environment Protection Authority Victoria, Australia (EPA Victoria).

** Portland Aluminium is a smelter operated by Alcoa of Australia Limited, with principal shareholders being Alcoa of Australia, Citic, and Marubeni. Portland Aluminium is located in Portland, in the state of Victoria, Australia. Point Henry is a smelter owned by Alcoa of Australia and is located in Geelong, also in Victoria.

† EPA Victoria is a statutory authority, established under an Act of the Victorian State Parliament, which has a charter to protect the Victorian environment.

†† Ausmelt Limited is a Victorian company providing innovative solutions for the metals and waste-treatment industries.

‡ CSIRO is a government-established science and research body.

‡‡ Interview with Helen Tope.

He is also representing Alcoa in a similar PFC reduction task force in Quebec, Canada. Tabereaux is that rare combination of a brilliant research scientist and a production manager who knows how to roll up his sleeves and make aluminum.

Back in 1995, Tabereaux and Dr. Jerry Marks, who managed Alcoa's Technical Center Analytical Division from 1986 to 2000, were among the select industry experts who participated with EPA representatives in an Aluminum Association "PFC Task Force" to reduce PFC emissions from US smelters. The Task Force provided direction and focus in the formulation of a new innovative voluntary emission reduction program with EPA. Under this partnership, each US aluminum company could set its own goal for reduction of PFCs based on the technology available at each location.

The EPA, as part of this voluntary partnership, agreed to provide assistance in conducting fundamental research and development (R&D) to discover the mechanism for PFC generation in pots, and to conduct PFC emission measurements at US smelters to develop more accurate equations to calculate PFC emissions at smelters based on different pot technologies. As part of the globalized effort, company and association efforts were initiated in Europe, Canada, Australia, and in other aluminum-producing regions.

Achieving the voluntary reduction in PFC emissions at Alcoa smelters involved teamwork. Many Alcoans played key roles in figuring out just how much PFC was being produced and how to eliminate its emissions. Jerry Marks and Ruth Roberts developed accurate analytical methods and actually measured PFC emissions from all Alcoa aluminum smelters. Charlie St. Clair and Gary Tarcy led an Alcoa technology team that developed and exploited a sophisticated alumina feed control algorithm and state-of-the-art pot hardware systems to reduce both the frequency and duration of anode effects at Alcoa smelters. And Pat Atkins, Greg Kraft, and Ken Martchek put together and administered an effective corporate strategy that ensured the success of the voluntary program with the EPA to reduce PFC emissions from all Alcoa smelters. The EPA's Sally Rand persuaded environmental regulatory authorities to give voluntary climate protection partnerships the time necessary to demonstrate success. Then she took the message back to the aluminum companies that voluntary reductions in greenhouse gases would assure freedom to pursue the most cost-effective options, rather than typical command and control imposed by regulators. "Do it yourself," she said, "before others tell you how to do it."

ENERGY-EFFICIENCY BENEFITS OF ALUMINUM *THAT DOES NOT NEED TO BE PAINTED*

" For starters, an unpainted plane can weigh hundreds of pounds less than its painted equivalent. Lighter aircraft naturally burn less fuel, with each pound of overall fleet-weight translating into thousands of dollars in fuel expenses each year. Add the fact that when paint chips or peels it increases an airplane's wind resistance, thus reducing its fuel efficiency, and it's clear that flying lighter, unpainted aircraft is a big help in our effort to control fuel expenses. And, since it takes more time to strip and repaint than to polish, not painting produces labor and materials cost savings as well. There is a safety benefit, too, since it's easier to find corrosion or dents when there's no paint covering an airplane's surface. "

Don Carty, Chairman and CEO, American Airlines, in American Way in-flight magazine, January 1, 2001

To reduce PFC emissions from aluminum smelting, Alcoa implemented a set of "best management practices." It educated employees on practices that reduce the frequency and duration of anode effects; supplied employees with the tools to monitor alumina concentrations; and held regular employee-involvement team meetings to help identify, develop, and implement anode effect, voltage, and energy reduction measures. Other efforts to reduce PFC emissions focused on "technical initiatives" to implement state-of-the-art technologies, such as computerized anode-effect suppression systems that reduce anode-effect duration and point-feed systems that control alumina feed.

Principal challenges included maintaining or improving production, energy, and current efficiencies at operating smelters during testing and full-scale implementation of efforts to reduce anode effects. In addition, improvement opportunities vary substantially for the four major technology subtypes for producing primary aluminum.[6] Similarly, smelting performance varies subsequently based on pot type and vintage even within one of these four technology types. Finally, Alcoa smelters are widely dispersed throughout the world based on their proximity to economic sources of electrical power, and it was a challenge to work among these variations and wide geographies.

In response, Alcoa encouraged significant technical sharing among its smelting personnel at facilities throughout the world. A technical lead team was commissioned to facilitate technology transfer. "Like technology" project teams were sponsored and supported on a worldwide basis. Alcoa personnel also worked through regional and international aluminum associations to share knowledge and leverage R&D efforts related to fundamental understanding of the nature of PFC emissions and related to PFC measurements.

Progress made by Alcoa has been impressive. Figure 1.5 summarizes the significant achievements made by Alcoa smelters over the past decade in reducing anode effects and subsequent PFC emissions.

Significant reductions occurred in all geographic regions as a direct result of project initiatives, testing, and implementation of best practices and technology improvements. Overall, Alcoa smelters reduced PFC emissions rates by 57% from 1990 to 2000. Worldwide, CO_2 equivalent emissions were reduced by about 9.9 million metric tons.

Alcoa smelters in Massena, New York, and Mt. Holly, South Carolina, have reduced PFC emissions to world-class benchmark levels as of year 2000. Highlighted in Table 1.1 are Alcoa smelters that achieved the highest percentage reduction in PFC emissions in the 1990s.

---

6   The most modern technology is the point-fed, center-worked, pre-bake cells (CWPB). Other reduction technologies used are the side-worked, pre-bake (SWPB) cells, vertical-stud Soderberg (VSS) cells, and horizontal-stud Soderberg (HSS) cells. DC currents are supplied to these cells ranging from about 60 k amperes in the older cells to 500 k amperes in the most modern cells. A series of reduction cells are aligned either end to end or side by side in rooms referred to as "potrooms." In recent years computer process control has been extended to most potlines. However, the more modern CWPB cells generally operate with higher electrical efficiencies, due to better design and to improved process control, than do older cell technologies. In addition, the modern cells can operate with advantages of higher productivity relative to manpower.

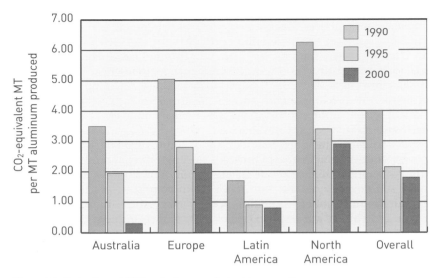

Figure 1.5 Reduction of PFC emissions made by Alcoa smelters

Smelter	Base year 1990	Actual 2000	Percentage reduction
Eastalco	18.83	1.55	92
Pt. Henry	4.34	0.53	88
Massena	0.67	0.10	85
Portland	3.01	0.45	83
Tennessee	4.19	0.80	81
Badin	2.42	0.46	81
Lauralco '93	0.30	0.06	80
Mt. Holly	1.02	0.20	80
San Ciprian	3.57	0.76	79

Table 1.1 Alcoa smelters achieving highest reduction in PFC emissions, 1990s

Aluminum's contribution to climate protection

Beyond aluminum production, the use of aluminum products in transportation—particularly in vehicle structures—significantly reduces CO_2 emissions as a direct result of weight reduction. According to the International Aluminium Institute (IAI), every pound (0.45 kg) of aluminum replaces two pounds (0.90 kg) of steel and saves 20 lb (9 kg) of CO_2 (by burning less fuel) over the lifetime of a vehicle.[7]

Ford Motor Company estimates that vehicle weight reduction with aluminum also reduces solid, waterborne, and airborne emissions by more than 13,000 lb (5,900 kg) over the life of the vehicle. The average 250 lb (113 kg) of aluminum in each vehicle produced in the USA (15 million per year) reduces CO_2 emissions by 75 billion lb (34 billion kg) over the projected life of the vehicles.[8] In addition, lightweight aluminum acts as a heat sink for the motors and all moving parts, reducing the need for cooling systems that would add weight, cost, and complexity.

Modern vehicles can further increase fuel efficiency by up to 8% for every 10% reduction in weight, by substituting aluminum for heavier metals. Fuel savings over the life of the vehicle more than offset the initial higher material costs to produce the vehicle; and the recycling of that aluminum later makes subsequent vehicles even more energy-efficient and affordable. Conservative estimates indicate savings can amount to 500–700 gallons (1,900–2,650 l) of gasoline over the lifetime of a vehicle, or $750–1,050 based on gas priced at $1.50 per gallon ($0.39 per liter).

Car manufactures have demonstrated that aluminum can replace steel with weight savings of up to 40%, resulting in a typical increase in fuel efficiency of 30%. The National Highway Safety Traffic Administration states that "Vehicle weight reduction is probably the most powerful technique for improving fuel economy."

And aluminum can make cars safer. Pound for pound, aluminum is up to two and a half times stronger than steel and, with today's advanced designs and honeycomb structures, can absorb crash energy without shattering.[9] The safety benefits of aluminum were realized in 1999 when the *Wall Street Journal* ranked the Audi A8, which employs an aluminum space frame and aluminum body panels, as the safest sedan in the world. In addition, occupants of vehicles struck by aluminum vehicles are safer because there is less inertia directed against their vehicle.

Increasing aluminum content also made cars more recyclable. Today, more than 90% of the aluminum content of a scrapped vehicle is recovered and recycled into useful products. In addition, more than 60% of the aluminum in today's new vehicles is sourced from recycled aluminum. Recent estimates by the IAI indicate that two-thirds of all of the aluminum ever produced is still in productive use and is leading us toward a more sustainable future.

---

7   The IAI study, which used peer-reviewed data, was compiled in accordance with International Organization for Standardization (ISO) standards.
8   Assuming that the aluminum replaces iron or steel, and based on 1999 model year production of 15 million vehicles: 15 million × 250 lb (113 kg) × 20 lb (9 kg) CO_2 savings per vehicle.
9   See www.autoaluminum.org/spr.htm.

Figure 1.6 Aluminum Honda Insight: the world's most fuel-efficient motor vehicle depends on lightweight aluminum

Other examples of aluminum-intensive cars include the Acura NSX, Audi A2 and BMW Z8, Ferrari Modena 360, Honda Insight (see Figure 1.6), and the Plymouth Prowler. The Audi A2 is the world's first high-volume, aluminum-intensive passenger car. The Oldsmobile Aurora has the highest aluminum content—480 lb (217 kg)—of any American production vehicle.[10] And almost every car has some aluminum. Through process improvements and technology innovations, and through increasingly valuable products and recycling, Alcoa is indeed positioning aluminum as the climate protection metal.

Another sector that benefits greatly from the use of aluminum is air transport. Lighter aircraft naturally burn less fuel, with each pound of overall fleet weight translating into thousand of dollars in fuel expenses each year. Since the 1930s, American Airlines has chosen an unpainted gleaming aluminum skin as its aircraft trademark.[11] American likes the crisp, no-nonsense look of polished aluminum "silver birds" (see Figure 1.7) but is quick to acknowledge that they have found numerous benefits to not having to maintain painted planes. In addition to the fuel efficiency lost from the added weight, when paint chips or peels it increases an airplane's wind resistance. There is a safety benefit, too, since it is easier to find corrosion or dents when there is no paint covering an airplane's surface.

10   The common definition of a production vehicle is more than 10,000 units per year.
11   *American Way*, January 1, 2001.

Figure 1.7 American Airlines Boeing-727 "silver bird"

Alcoa time-line

6000 BP	Copper is first metal smelted for tools.
5500 BP	Copper and tin smelted into bronze for weapons, tools and armor.
1808	Sir Humphry Davy (Britain) discovers and names aluminium.
1821	Bauxite, the most commonly mined aluminum-bearing ore, first discovered in the French district of Les Baux, after which it was named.
1845	Friedrich Wöhler (Germany) establishes the specific gravity (density) of aluminum.
1854	Henri Sainte-Claire Deville (France) creates the first commercial aluminum production process. The price of aluminum—previously higher than that of gold and platinum—drops by 90% over the following ten years.
1886	Charles Martin Hall (USA, founder of Alcoa) and Paul Louis Toussaint Héroult (France) separately and simultaneously invent a new electrolytic process, which is the basis for all aluminum production today.
1888	Alcoa founded as the Pittsburgh Reduction Company in 1888 by Charles Martin Hall in Pittsburgh, Pennsylvania, USA.
1904	Large-scale aluminum recycling begins in Chicago and Cleveland.
1957	First aluminum cans produced.
1980	Report by R.A. Rasmussen reveals the manufacture of aluminum to be a primary source of PFCs.
1994	The Audi A8 becomes the first production car model with an all-aluminum body.
1995	Voluntary Aluminum Industrial Partnership for PFC Reductions is launched.
2001	Alcoa Portland SPL process developed.

2
Aviation Partners
The future is on the wing*

How a "dream team" of retired aeronautical engineers and test pilots challenged the conventional wisdom of aerodynamics to develop a new concept for Blended Winglet technology which increases the fuel efficiency of aircraft up to an astonishing 7%, then formed a joint venture with Boeing to retrofit the global aircraft fleet.

Thrust, lift, weight, drag, stability, and control are the technical obsession of all aerospace engineers, from the ancients who first studied the wings of birds to modern computer-trained engineers.

Although he would not have thought about them in those terms, Ibn Firnas was concerned about those challenges of flight when he built the first flying machine in AD 875. Ibn Firnas watched birds flying and studied their wings carefully. But he failed to notice how birds use their tails when they land; and he ended up crashing his invention into a mountain outside Cordoba, Spain. He told the spectators that he had forgotten to add a tail to his machine, like a bird's tail.[1]

Leonardo da Vinci faced the same challenges of flight. Like Ibn Firnas, he studied birds and bats to learn the secret of their flight. He also dissected them. Among other things, he observed that the inner part of natural wings moved more slowly than the outer part, and concluded that the function of the inner part was to sustain rather than to push forward. In imitation of the birds and bats, he developed working gliders with fixed inner sections and mobile outer sections.[2]

* The authors are grateful for interviews and supplementary assistance by the following engineers and managers at Aviation Partners: Joe Clark, Maggie Clark, Dick Friel, Kim Frinell, Louis "Bernie" Gratzer, Robert T. Lamson, W.S. "Bill" Lieberman, Ted Lomax, and R.L. "Dick" Sears; by the following engineers and managers at Aviation Partners Boeing: Kevin Bartelson, Sheldon Best, Mike Stowell, and Jay Inman; and by President Clay Lacy of Clay Lacy Aviation and President Borge Boeskov of Boeing Business Jet.
1 See www.angelfire.com/realm/bodhisattva/flyers.html.
2 Information posted on www.angelfire.com/electronic/awakening101/leonardo.html by Museo Della Scienza e Della Tecnica.

In 1638 in Turkey, Hezarfen Ahmet Celbei built a wing apparatus inspired by da Vinci's design, and launched it from the 183-foot-tall (55 m) Galata Tower near the Bosphorus in Istanbul. Celbei had watched an eagle in flight, and he made adjustments to the da Vinci design based on that great bird. His flight was successful.[3]

In his 1889 book, *Bird Flight as a Basis of Aviation*, Otto Lilienthal wrote "we are forced to consider the flying apparatus of the bird as a most ingenious and perfect mechanism." Later, when the Wright brothers were trying to figure out how to control an airplane in the sky, they too looked at birds. Wilbur Wright noted that birds twist their wing tips to stay in control; and he built a box kite with wings that could be twisted in opposite directions, based on this idea, to make it bank and turn. The Wright brothers called this principle "wing warping," and it has inspired hang-glider designers ever since.

"It doesn't break the laws of physics, but it does bend them rather beautifully," says a Boeing advertisement praising improved performance and fuel economy.

High-tech engineers are still thinking about these same challenges of flight, and are still inspired by the design of bird wings. Today, aerodynamic engineers concentrate their efforts to enhance performance on lift, weight, and drag. Drag is the

BLENDED WINGLET CHARACTERISTICS

The Aviation Partners high-aspect-ratio Blended Winglet is a significant departure from the conventional winglet design. It features a large radius and a smooth variation in chord in the transition section. It is engineered to provide optimal performance, limited by wing structural strength and dynamic characteristics. This allows optimum aerodynamic loading and avoids drag-producing vortex concentrations. Correctly designed Blended Winglets have demonstrated much smaller wingtip vortices than straight wing aircraft or conventional winglet systems with angular transitions.

"Blended Winglets™" are gracefully curved wingtip extensions that enhance the aerodynamic efficiency and flight characteristics of aircraft. Aircraft equipped with Blended Winglet systems are environmentally superior because they can fly farther on the same amount of fuel and they are quieter on take-off. In addition, Blended Winglets allow take-off at higher gross weight, which is particularly beneficial at high-altitude airfields and during hot weather. Blended Winglet-equipped aircraft require less power to do the same job, resulting in lower engine maintenance. Owners enjoy increased range, performance, and cargo payload while significantly reducing greenhouse gas and toxic emissions. Reduced noise helps satisfy increasingly strict standards, particularly in Europe. Blended Winglet technology has changed the way the aviation world thinks about optimizing wing efficiency.

Aviation Partners Incorporated (API) introduced its patented Blended Winglet Systems in 1991 as a performance enhancement to the legendary Gulfstream II business jet. Since then, it has been adapted to an increasing number of airplane types (see Figure 2.1). Blended Winglet technology is equally effective on virtually any make or model of business or commercial aircraft in service today—suggesting limitless market and environmental potential as fuel use and emissions become more important.

---

3   See the chapter on Hezarfen Ahmet Celbei in Evliya Celebi's book, *Seyahatname* (*Book of Travel*) published in eight volumes 1896–1928, as posted on www.angelfire.com/electronic/zennun/celebi.html.

Figure 2.1 Boeing business jet with Aviation Partners Blended Winglets

aeronautical term describing resistance to airflow. An aerodynamic shape has less drag because it is streamlined and can cut through the air more efficiently. Reducing drag allows aircraft to move forward with less thrust, and so reduces fuel consumption. Conversely, for every increase in drag, there must be a corresponding increase in power and fuel consumption.

To put this into perspective, a Boeing 747 requires a gallon of fuel for every 6.7 lb (3.8 l for 3 kg) of its flying weight. As a result, it has to haul up to 300,000 lb (136,000 kg) of fuel. A more aerodynamic design allows an airplane to carry less fuel and more passengers or cargo, or to have the capacity for longer non-stop flights which add an important margin of safety and usually command higher ticket prices. For people in the aviation business, it is a simple equation: make planes more efficient by reducing drag, and you generate money on cargo and passengers and save on fuel at the same time. It is a central goal in the industry.

For those concerned about the environment, improved aerodynamics and the resulting savings of fuel reduces nitric oxide air pollution, climate-changing carbon dioxide emissions, and acid-rain-causing sulfur dioxide emissions. It also reduces the noise footprint on the ground because aircraft with less weight and drag climb out of airports more quickly.

A brief history of winglets

"Everyone knows that a wing creates a vortex at its tip, which creates drag," explained Dr. Bernie Gratzer when we interviewed him at Aviation Partners about the company's Blended Winglet technology. Without sounding condescending, Gratzer continued his lesson. "Drag, we know, is the enemy of flight, slowing the airplane as it moves through the air. Reduce drag, and you increase lift and velocity."

Gratzer is a young 81 years of age and is the only member of the engineering team at Aviation Partners who wears a tie. His wispy, white hair makes him look like the professor he was after he left Boeing, where he had served as the chief aeronautical engineer. He has a self-deprecating way of smiling and chuckling, almost to himself. Gratzer has the clear eyes and the twinkle that we would come to recognize as the Aviation Partners team trademark.

In order to understand winglets, one must understand wings. Air flowing over the upper curved surface of a wing accelerates when it is forced to travel further than the air flowing directly under the lower wing surface. The difference in pressure between the upper and lower wing surfaces generates necessary lift.

However, high pressure on the undersurface of the wing causes some air to escape at the wingtip, forming a powerful helical wake vortex (see Figure 2.2) which reduces the available lift. The turbulence that results wastes energy by diminishing lift and increasing drag. The turbulence of the wingtip vortex is strong enough to flip airplanes flying too close behind one another and it is a critical reason for the stringent requirements for separation distances between aircraft, particularly at take-off. The margin of safety for any aircraft decreases as the load of fuel, passengers, and cargo approaches the maximum aircraft rating because the aircraft becomes less able to accelerate or to execute the tightest maneuvers.

For more than a century, physicists and aviation engineers have struggled to reduce the effect of wingtip vortex and the drag it exerts. One method is to make the wing longer, increasing its aspect ratio. The aspect ratio of a non-tapered wing is its span divided by its chord. The span is the distance from wingtip to wingtip, and the chord is the distance from the leading edge to the trailing edge. A wing with high aspect ratio is more tapered and loses proportionally less energy to the wingtip vortex than a more stubby wing does. But longer wings pay a penalty in weight from the added materials needed to support increased bending.

Another way to reduce drag is to use winglets, which provide the effect of increased aspect ratio, but with less weight than a longer wingspan. Winglets are small, up-swept attachments to the aircraft wingtips that extend the trailing edge and relocate and reduce the strength of the vortex, thereby reducing drag.[4] Wing-

---

4 Winglets are wingtip devices that primarily reduce induced drag, which along with other lift-dependent drag forces makes up almost 50% of airplane total drag. This can be expressed by the formula:

$$\text{Lift-dependent drag} \quad = \quad \frac{\text{Lift}^2}{\text{Span}^2 (p\,q\,e)}$$

In this equation the dynamic pressure, q, equals one-half the product of air density and airplane velocity squared. The Oswald efficiency factor, e, accounts for lift-depen-

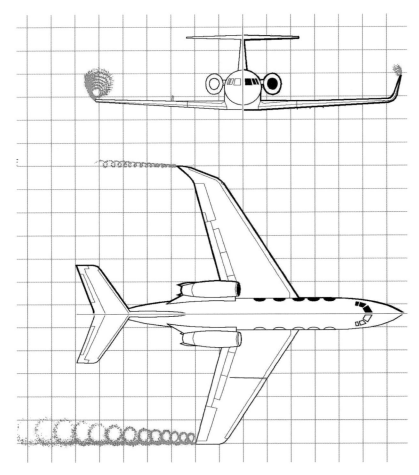

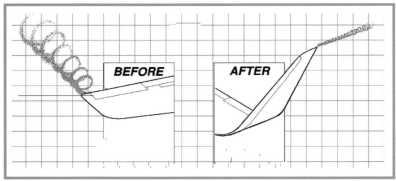

Figure 2.2 Aviation Partners Blended Winglet reduces vortex

lets are not new. "The original idea of winglets has been around a long time," explained Gratzer. In 1897, F.W. Lanchester patented a vertical wing-end plate. But there was scant interest until 1974, when NASA engineer Richard T. Whitcomb published the results of theoretical and wind-tunnel studies predicting potential fuel savings of 4–5% from the technology. NASA and the US Air Force began flight-testing in 1979 and confirmed Whitcomb's estimates on a KC-135 equipped with winglets (provided under contract from Boeing). The US Air Force, however, did not proceed farther with winglet technology after initial testing.[5]

In 1981, McDonnell Douglas tested the NASA designs further, and later used them on MD-11 and C-17 Globemaster III aircraft. Around 1983, Learjet became the first company to install winglets on a business jet. In the 1970s, Airbus had been first to equip a production airliner with small winglets, but the company changed back to a conventional wing for the A330/A340. Later, the Ilyushin Il-96-300 also employed conventional winglets; and Valsan obtained FAA (Federal Aviation Administration) approval to retrofit winglets to the 727.[6]

But these early efforts used large, boxy winglet designs and achieved only marginal improvements in fuel efficiency. By 1990, most aviation experts had dismissed or opposed winglets because they were sized improperly, shaped poorly, and did not work very well.[7]

Never too old to do the impossible

After nearly a hundred years—and the best efforts of NASA and the major major aircraft manufacturers—the aerodynamics of wings and winglets had proven too complicated to improve with winglets. Because the giants of aerospace hadn't been able to do it, many smart people assumed that it couldn't be done. But perhaps this

dent components other than induced drag. The effectiveness of winglets stems primarily from their ability to increase equivalent span, but they also improve efficiency by optimizing the aerodynamic loading of the lifting system. The API Blended Winglet design incorporates a smoothly curved transition between the wing and the winglet. A conventional winglet in contrast, has an abrupt connection to the wing with undesirable flow characteristics that limit winglet performance. The Blended Winglet with high aspect ratio promotes full realization of winglet drag reduction potential while minimizing friction and interference drag factors that detract from the primary benefits. Although a basic wingspan increase will also reduce drag, the fundamental advantage of winglets is that they produce substantially less wing-bending moment than a planar wing. Winglets require less structural weight for the same equivalent increase in span. Thus, for an optimized winglet design application, the benefits appear as reduced drag as well as reduced span and structural weight.

5   Flight tests on the KC-135 indicated that winglets resulted in about 5% fuel savings—45 million gallons (171 million liters) per year—but the US Air Force chose a Fuel Savings Advisory/Cockpit Avionics Systems upgrade instead (Ed Davies, "Winging It: Boeing Welcomes the Winglet," *Airways*, April 2001).

6   *Ibid.*

7   Bob Shane, "Modern Winglets: A Lift for Commercial Airlines," *Airliners* 72 (November/December 2001): 1-4; Fred George, "Getting a Lift Out of Winglets," *Business and Commercial Aviation International*, February 1998.

was really like the economist who was so captivated by his assumptions that the market was perfectly rational that, when he saw a $100 bill lying on the sidewalk, he walked on by, assuming that someone else would have picked it up had it been a *real* $100 bill.

On the other hand, if NASA and the other giants of aerospace couldn't improve on the wing design or the winglet, then who in the world would be audacious enough to try? And would anyone take them seriously?

It turned out that a Montana industrialist named Dennis Washington would try, and his interest would ultimately set in motion a series of events that would pull Gratzer and several others out of retirement and give them a chance to do what no one else had ever been able to do before.

Dennis Washington became an industrialist when he bought the Anaconda copper mine in Montana, back at a time when the price of copper was so low that the mining machinery was worth more than the ore itself. Washington bought the mine with a mind to sell off its trucks and equipment. But the price of copper went up and up, and he ended up keeping the mine. He also ended up making a lot of money.

Washington was flying a Gulfstream II at the time, and he enjoyed the convenience and speed of his own private jet. He was frustrated, however, that his Gulfstream was range-restricted on coast-to-coast flights, and so when the new Gulfstream III came out he was a bit envious. It had a more stylish design and a larger range. But Washington didn't want to spend millions more for a newer-model Gulfstream. He wondered why there wasn't a less expensive way to modify his older Gulfstream to make it look more stylish and to increase its range—to make it more like its expensive sibling, the Gulfstream III.

Washington called his friend Joe Clark. Clark is not an engineer or a former test pilot. He has, however, been hooked on airplanes ever since his first flight on a Learjet in 1964. In 1965, Clark founded Jet Air, the first Learjet distributorship in the Northwest. Jet Air started out with only one jet and only one customer, a company his father operated; but it was the first golden domino in a lifetime of successful ventures. It grew quickly into a profitable business with a sales territory covering Washington, Oregon, Alaska, and all of Canada. After Jet Air, Clark and Milt Kuolt co-founded Horizon Air, a Seattle-based regional airline that was later sold to Alaska Airlines. Clark also founded Avstar, a global sales network that took advantage of good deals on military-surplus training jets and sold them to private companies and individuals.

Washington asked Clark if his Gulfstream II could be modified to look more like the Gulfstream III and to achieve greater flight range. Clark considered the problem. If he were able to dramatically improve the aerodynamic efficiency of the Gulfstream II's wing, he might be able to solve the problem posed by Washington and give him what he wanted. But Clark knew that airplane wings had had the same basic design since the 1950s, and he was familiar with the general consensus among aerospace engineers in 1990 that there were few, if any, possibilities to improve on the standard wing. A different man might have shrugged off his friend's request and gotten on with his busy life.

But Clark's curiosity was piqued. He had been successful long enough to know that the first rule in business is to know your own strengths and weaknesses. Clark recognized that he wasn't the man to say that the wing could, or couldn't, be improved. And this new aviation puzzle intrigued him. He decided to accept Washington's challenge, and see if he could put together a team to figure out how to substantially reduce wing drag while increasing the aesthetic appeal of the Gulfstream II aircraft.

Clark started by calling his friend Kim Frinell, and together they studied the technical and financial feasibility for a one-time Supplemental Type Certificate (STC) to develop a winglet to make a GII more efficient. When the study concluded that the modification would provide significant improvements in aircraft performance, Clark decided to form a new company to develop the project.

Clark called a meeting at Frinell's house, and included Bill Lieberman, Dr. Bernie Gratzer, Bob Stoeklin, and Peter Jennings on the invitation list, along with Dennis Washington. The team Clark assembled had the necessary engineering experience; or at least he hoped it did. And Clark, Frinell, and Lieberman had the key business connections at Gulfstream Aerospace in Savannah that would be needed to make the effort a reality, he hoped. By the end of the evening, the team had a business plan and enough optimism for Washington to agree to fund the venture.

Bernie Gratzer and Joe Clark with Spiroid and Blended Winglets

The team prepares for corporate lift-off

The team was rolling; but it still faced some major hurdles if it was going to accomplish its ambitious goal. Surpassing one hurdle would not be enough. Only by running the full race could the team redesign a major industry. And even practical needs, such as getting data from proprietary sources, could prove to be an insurmountable hurdle.

The engineering challenge was to test the Blended Winglet on the Gulfstream II. When Gulfstream designers balked at providing engineering data and engineering experience, Bill used the back doors he had cultivated as a salesman and got management override. Dr. Gratzer and the team collaborated on the design; a Seattle manufacturer of competition rowing shells fabricated the prototype proof-of-concept winglets; and they were in business. Washington provided his Gulfstream II for flight tests. Dick Sears, Frinell, and Lieberman conducted "before and after" Blended Winglet tests on the Gulfstream II to confirm Gratzer's predictions.

A second hurdle was to get Gulfstream II owners to buy the Blended Winglets. Clay Lacy and Dick Friel were key in accomplishing this step, and their personalities show why. When we came to interview the team for this chapter, Clark invited us to join him at Randy's Diner, a short stretch from Boeing Field. The diner was in a mixed-use area near Boeing field, and had the feel of place where a working man or woman could come for a lunch that would be the same every day. The design was classic: Formica® table tops in booths that were lined up around the outside edge of the diner, and a counter in the interior for those eating alone. There was an added touch: models of Boeing and other aircraft suspended from the ceiling. The cast of regular customers greeted each other by name, and our saucy waitresses knew just what Jo and Clay wanted for lunch.

Clay Lacy met us there. Lacy is a big, genial man in his late sixties who has a winning smile and the soft drawl of his native Kansas. Lacy tells us about a day well spent, going to school with his friend's four-year-old daughter. The twinkle in his eyes shows his obvious delight, and you can see that Lacy is a terrific grandfather. Just about the whole of Lacy's life has been dedicated to flying. He built his first gas-powered model airplane at age eight, and by the time he started flying, at age 12, he was completely enamored with aviation. One of his first jobs was flying "right seat" on Douglas DC3s, back in 1952 as United's youngest pilot. By the time he retired, he was United's number one pilot in seniority, flying 747-400s from Los Angeles to Sydney, Australia. With over 50,000 hours of flying under his belt, in every kind of plane, from Mustang P-51s to the 747s, Lacy has probably logged more hours than any pilot on Earth.

WINGLET BENEFITS

- Reduced fuel consumption and emissions
- Reduced engine maintenance
- Increased flight range and payload
- Greater margins of safety
- Improved handling and better stability
- Faster climb to altitude
- Higher initial cruise altitude
- Quieter skies
- Sleek and modern profile that modernizes an operator's fleet
- Higher resale value (style and performance)

Lacy has flown all over the world, from Beijing to Moscow. In 1968, he founded Clay Lacy Aviation and built it into what many consider to be the premier West Coast business air-charter company. He was one of the first Gulfstream II owners to adopt Aviation Partners Incorporated (API) Blended Winglet technology, and he set more world records in his 22-year-old Gulfstream IISP than in any other aircraft he's owned or flown. In June of 1995, the winglets helped Lacy and Clark set a new Gulfstream climb record and a world speed record from Los Angeles to Le Bourget, Paris, on the way to the Paris Air Show. With an elapsed time of 10 hours and 36 minutes, and an average speed of 531.8 miles per hour (mph) (850 km) over the 5,638 mile (9,021 km) great circle route, the Gulfstream IISP achieved an 83 mph (133 km) gain over the previous record of 448 mph (717 km). On the return home, Lacy and Clark established a new world speed record from Moscow to Los Angeles. Ultimately, Lacy's world records and the publicity around them, as well as his success with the Blended Winglet technology, helped him and his colleagues to promote the winglets.

The characters of the team

So much about the success of the winglet technology came down to the characters of the individuals involved. Nothing about their personalities was to be taken for granted; and their coming-together was not inevitable. In fact, it seemed quite far-fetched at the beginning. Far from an ordinary corporate gathering, it was more like the elaboration of a film script. It was like the movie *Space Cowboys*, where Clint Eastwood, Tommy Lee Jones, Donald Sutherland, and James Garner are retired test pilots who come out of retirement to fly a dangerous mission to disarm nuclear weapons on an old Soviet space platform.

Even the API's office space came more out of a Hollywood script than from Main Street. At their humble beginnings, with six engineers working around one conference table, Clark housed Aviation Partners' offices in the Flight Center FBO (Fixed Base Operation) at Boeing Field. The design engineers for Aviation Partners and its offshoot, Aviation Partners Boeing, now occupy what used to be the restaurant "Blue Max" in the original terminal building at Boeing Field. Everyone keeps an eye on the long runway outside the window, often interrupting conversations to comment on whose plane is taking off or landing. "That's Clay Lacy in his Learjet;" or "that plane is flying our winglets." Despite Aviation Partners' commercial successes, the office still has the feeling of the start-up that it was not so long ago.

Dick Sears is a former test engineer who was head of flight safety during his last ten years at Boeing. In addition to heading API's flight test program, Sears is also program manager on the new Hawker 800 Blended Winglet project. Sears is a handsome guy, with a full head of white hair.

Lamson is 88 years old, and has the trademark twinkle of the team at API. His long career includes many breakthroughs. His first career was as an engineering test pilot for Boeing back in the forties and fifties, when he won the prestigious Octave Chanute award. He did the experimental test flights of nine different types of plane including the B-50 bomber and the XF 8B1. From 1949 to 1951, Lamson was the chief test pilot on the 377 Stratocruiser. After serving as a test pilot with

THE DREAM TEAM

- **Joe Clark**, aviation entrepreneur
- **Dennis Washington**, Montana industrialist
- **Dick Friel**, sales and marketing master
- **Dr. Louis "Bernie" Gratzer**, chief aerodynamicist and designer
- **Bill Lieberman**, program manager
- **Robert Lamson**, composite guru
- **Dick Sears**, Hawker 800 program manager
- **Kim Frinell**, program manager director of engineering

Boeing, followed by a second career as an aviation technical consultant, Lamson launched a third career in composite technology, reviving his pioneering interest in organic chemistry. Many years earlier, in 1935, he published the seminal technical paper on synthetic plastics and their application to the modern airplane.

In 1965, Lamson built the world's first high-performance pressurized sailplane out of composite materials. When he was honored, in 2001, as a Pathfinder at the Museum of Flight in Seattle, Washington, the award carried the following notation: "This would have been a remarkable technical achievement had [the sailplane] been built by a highly skilled team on a factory floor; that it was hand-built by one man in his own shop marks it as the work of a true Pathfinder." The sailplane, with a wingspan of 70 ft (21 m), now hangs in the Museum of Flight's Great Gallery.

Lamson's role in the API Blended Winglet program is in composite materials. He supervised the construction of all prototype winglets used on API's test flights. Lamson deflects credit to others, "It is the design team and the flight test program guys who are the real heroes here."

Along with Lamson and Sears, the team includes Gratzer, Lieberman, and Frinnel. Except for Frinnel, all are in their seventies or eighties, and all are long retired from Boeing. But they are obviously not done working. Most of them still come into the API office, and for many of them this is not their only job. Several have consulting businesses as well. Clark explains,

> These guys have done it all. They grew up in an era when the best in the field knew everything about the planes they designed and flew; they could see the big picture, and they trusted their intuition. Today, the new generations are far more specialized, and only see the part of the problem they are assigned to solve. It's hard to optimize a plane when you only know part of it.

Lieberman, after a stint as a commander in the US Navy, joined Boeing as chief flight test engineer. At Boeing, he was responsible for flight tests and FAA certification of many planes. At one point, Lieberman took ten years away from aircraft design to sail his sloop *Nomad* about 35,000 miles (56,000 km) around the world. Lieberman also runs a consulting company that had worked for Gulfstream on certification. Like other members of the API team, Lieberman is a confident and respected engineer with a record of success.

Frinell, the youngster of the group, alludes to a checkered past working for Boeing. "Lieberman hired me four times for Boeing flight test programs and only fired me once," he says. Frinnel has two first flights to his credit—the DC 9-30 and the 737-200.

In conversation, Lamson talks frequently about the team. It is clear he not only knows planes, he also knows people. "I'm a lucky man," he says: "I get to work with the best in the business. World speed record holders. Legendary test pilots. The most talented aerodynamic engineers. The smartest folks in the history of the business. Bernie Gratzer is a real engineering genius with the eye of Leonardo da Vinci—he can draw a winglet with performance *and* style."

We were soon introduced to another character: API's Vice President of Sales and Marketing, Dick Friel. Frinnel describes Friel as "a group of people in one pair of

Blended Winglet being installed on a Gulfstream

pants, with the humor of a dozen comedians." Unlike other members of the team, Friel looks very Hollywood in his elegant Saville Row shirts, with French cuff and solid gold cufflinks, his yellow yachting tie, and his suspenders. Friel hired Clark when he was VP of Marketing and Sales for Gates Learjet; and after that Friel took over marketing management of a home improvement retail corporation and then started his own advertising agency. Friel came back to aviation when Clark started API and needed a marketing genius. "I just closed my company and came to work for Joe. I knew he was on to something really good," says Friel.

This top-gun marketing and sales wizard is credited with creating the award-winning Blended Winglets advertising and public relations campaign. In it, he translated the advanced technology terminology of the Blended Winglets into a "top-of-mind" awareness that moved the technology so well that it dominated the marketplace. "The bottom line," says Friel, "is that our winglets let you fly higher, faster, and farther. They make an old airplane look brand new again. They're sexy!" Friel is also an actor who has made more than 100 commercials and acted in several movies, including one with John Wayne. In his spare time, Dick serves on the Marketing and Capital Campaign Committee of the Museum of Flight in Seattle. Friel's experience and character was key to moving the winglets.

The hurdles keep coming

Once the team had jumped the second hurdle by getting Gulfstream II owners to buy the Blended Winglets, the third hurdle presented itself. Despite success with the Gulfstream II program, API faced daunting challenges in persuading Boeing to adopt Blended Winglet systems on its business jets. Boeing aerodynamic and structural engineers believed that the additional weight and drag of the winglets would cancel out the benefits.

But Boeing Business Jets (BBJ) President Borge Boeskov had customers who said that the Blended Winglets looked modern and sexy, and those customers were already asking for winglets. Compared to the small, angular, traditional winglets you often see on commercial and business aircraft, Blended Winglets are noticeably taller and feature a smooth curvature transition between wing and winglet. Boeskov was the best Boeing salesman ever, and he trusted his instinct that something had to be done to distinguish the BBJ from its commercial version, the basic Boeing 737.

Clark cut a deal with Boeskov at the 1997 Paris Air show to design and build winglets for BBJ—on the condition that they would become standard equipment if successful. But when Boeskov put in a request to Boeing for a test plane to bolt the winglets onto, he was told that none were available. Small obstacles such as that, however, just make life more entertaining for him. Boeskov is a Danish immigrant, by way of Iceland, who worked his way from earning $25 a week at a seed store to presiding over one of the most successful executive aircraft companies in history. "Easily discouraged" is not a term anyone would use to describe Boeskov. Earlier in his career, Boeskov had worked in the European sales division at Boeing and had met people over at Hapag Lloyd, a German company that owned a charter/tour fleet of 737s. Hapag Lloyd agreed to lend him one of their jets for six weeks, provided that he let them keep the winglets if they worked. The deal was set with a handshake.

But the Boeing engineers remained pessimistic. Bob Lamson remembers the pessimism that came from a particularly knowledgeable colleague. "My next door neighbor is a retired Vice President of Boeing. In his early years he designed and built the Boeing wind tunnel and is a great believer in wind tunnel testing. One day, in an informal conversation, he told me he suspected that our calculations were in error and our performance enhancement was too optimistic." Imagine being told by the inventor of the wind tunnel that your plans were doomed! A person with less confidence might not have had the courage to continue.

Dick Sears adds his own story about Boeing's pessimism:

> I remember when we were getting ready for the test with Boeing, and needed to visit the plant to go over the wing strength on the Boeing Business Jet to make sure we could fit our product on properly. That day a young engineer met me at the security gate to escort me to the wing division, which is near the back of the sprawling plant.

The 73-year-old Sears has the team's trademark sparkle in his eyes as he continues to tell his story.

During our long walk, this young man got to talking, and told me he didn't think our winglets would pass the test. He said that the Boeing engineers had calculated that our systems might require an extra 2,000 pounds [900 kg] of weight just to strengthen the wings to accommodate the Blended Winglets. I quietly told the young engineer that we'd calculated a total weight of between 200 and 300 pounds [90–135 kg].[8] He just shook his head. He clearly didn't believe me.

When API rolled out the borrowed Boeing 737-800, fitted with their Blended Winglets for flight testing, a Boeing aerodynamics engineer boasted that if API achieved even a 2% performance improvement then he would fire all of his engineers and would quit his job.

But the API team remained confident, and with good reason: they were right. At the test flight, Boeing confirmed a 5% improvement. To their credit, the Boeing engineers quickly got on board and are now among the strongest advocates of Blended Winglet technology.

Boeing ultimately formed the Aviation Partners Boeing Corporation for application of Blended Winglets to Boeing business and commercial jets. The joint venture has the advantage of providing full access to Boeing aircraft design specifications and engineering to the team, drastically simplifying evaluation of winglet configuration. And each application to the FAA for new winglet configurations builds on previous physics, engineering, testing, and flight experience with other Boeing aircraft—speeding approval.

However, the marketing challenges remain daunting. Engineers and pilots steeped in the old physics and technology remain skeptical about each new application. The benefits of reducing noise and local air pollution are undervalued and under-priced, and regulations on carbon dioxide emissions are not yet implemented globally. The benefits of safety from reduced wake turbulence at airports are neglected as well because "it would be hard to justify a technology that will 'benefit the guy behind you.' "[9]

But problems such as this have only been one more hurdle for the team to overcome. And they have had much on their side. In 1999, for example, Joe Gullion, President of Boeing Airplane Services, said, "With thousands of airplanes as potential candidates for winglets, we see this as a great retrofit option for our customers and a source for profitable growth for both the airlines and Boeing."[10]

API faced other challenges while they were trying to set up the joint venture with Boeing Business Jets. They had to bridge multiple cultures among engineers and business colleagues. "Many of the young engineers don't believe anything that isn't in their computer program," says Clark. "While the computer is a fantastic tool, if you're not careful it also can limit your thinking, especially when you are looking for a breakthrough, like we were. Our guys still know how to use a pencil,

8 Winglet designs for retrofit must accommodate the existing wing shape and structure for each aircraft, but can choose height, cant, toe, twist, taper, and transitional curvature. Cant is the angle the winglet is bent from the vertical. Toe is the angle of the winglet surfaces relative to the direction of airflow. Twist is the rotation of the winglet as seen from above, and taper is the rate of change in the thickness of the winglet airfoil surfaces.
9 "News Scan: Creating a Wake Vortex," *Scientific American*, February 2002: 16.
10 From Aviation Partners files.

a piece of paper, a formula, a slide rule, and the seat of their pants. Let me show you the first model of our elegant winglet . . ." Clark recovers from his office a winglet cut-out glued on one end to a piece of paper and propped up on the other end with a toothpick. Formulas and critical angles are penciled in. Somehow it seems impressive.

And it is impressive. API's Blended Winglet technology has demonstrated significant fuel savings on every demonstration airframe tested, and it should be suitable for virtually any aircraft. Figure 2.3 shows the performance benefits on Boeing 737s.

737-700/-800/-900:

With Optional New Technology Winglets

Winglet Performance Benefits

	737-700	737-800	737-900
Lower fuel consumption			
500 nmi	-2.4%	-2.5%	-2.2%
1,000 nmi	- 3.3%	-3.4%	-2.9%
1,500 nmi	-3.5%	-3.5%	-3.4%
Increased payload-range			
Design range increase	+115 nmi	+130 nmi	+55 nmi
Payload capability increase (fixed range)			
Fuel capability limit	+5,500 lb	+6,000 lb	+5,550 lb
Maximum takeoff weight limit	+1,500 lb	+1,000 lb	+1,000 lb
Improved takeoff performance			
Engine	CFM56-7B24	CFM56-7B27	CFM56-7B27
High/hot takeoff weight increase			
Denver, 30°C	+ 5,400 lb	+4,200 lb	+3,800 lb
Obstacle-limited takeoff weight increase			
Close obstacle (50 ft high, 300 ft out)	+2,200 lb	+1,800 lb	+1,300 lb
Distant obstacle (500 ft high, 8,000 ft out)	+3,400 lb	+3,600 lb	+3,800 lb
Reduced certification noise			
Takeoff noise reduction at cutback	-0.5 to -1.0 EPNdB*	-0.5 to -1.0 EPNdB	-0.5 to -1.0 EPNdB*

* Estimated performance for the 737-700 and 737-900

Range Capability

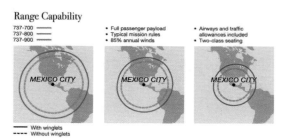

737-700 ——
737-800 ——
737-900 ——

• Full passenger payload
• Typical mission rules
• 85% annual winds

• Airways and traffic allowances included
• Two-class seating

—— With winglets
- - - Without winglets

The Boeing Company
Commercial Airplanes
Marketing
P.O. Box 3707
Seattle, WA 98124-2207

www.boeing.com

Figure 2.3 Winglet fuel savings, increased range and payload on Boeing 737s

Over 10,500 commercial Boeing airplanes are in service today. The Aviation Partners Boeing (APB) business strategy includes winglets for Boeing 747 and 767 airframes, including 747-200 freighters and the 747-400X passenger jets. "Induced drag and lift-dependent drag is responsible for almost half of the total drag of an airplane. We are able to reduce induced drag 12–18% or more with Blended Winglets and this translates into a reduction in overall drag of about 5–8%," says API's Chief Aerodynamicist Dr. Gratzer.

AIRCRAFT LEASING COMPANIES OPERATING WITH API BLENDED WINGLETS

- Air Berlin
- Air Europa
- American Trans Air
- Boullioun Aviation Services
- COPA
- GATX Capital Corporation
- GECAS
- Hainan Airlines
- Hapag-Lloyd
- ILFC
- Kenya Airways
- Pegasus Airlines
- Polynesian Airways
- Qantas Airways
- South African Airways
- Tombo Aviation
- Travel Servis
- VARIG
- Virgin Blue

The sky's the limit

You can't stop a team that is really on a roll—and Clark's team of retired engineers and test pilots has turned out to be the "dream team." Aviation Partners considers the extraordinary energy-efficiency benefits of Blended Winglet technology to be just the beginning. API's Blended Winglets are rapidly being transferred beyond their beginnings on the Gulfstream II business jet.

Soon, Blended Winglet technology will be introduced to dramatically improve the range and performance of the Raytheon Hawker 800 series business jet. In the commercial airline world, APB has brought Blended Winglet technology to next-generation series Boeing 737-800s and 700s. By mid-2003, this performance-enhancing technology will be available as a retrofit for Boeing 737 Classic series aircraft, including the 737-300/400/500. And APB's business strategy includes Blended Winglet systems for Boeing 747 and 767 series airframes.

Future products are in the works as well. Following the Gulfstream II winglet STC, Clark made good on his promise to flight-test Gratzer's closed-loop Spiroid winglet design (see Figure 2.5). Those tests on a high-speed jet validated a new concept, and indicated a 10% reduction in drag. Overall, they may be 40–50% more efficient than Blended Winglets. So far, however, only Lockheed's Skunkworks division has shown interest in this radical new performance-enhancing design for the U2.

Advanced-generation Blended Winglets with controllable twist and toe angle could be optimized in flight. One day, aerospace designers will likely integrate Blended Winglet and Spiroid technology into the design of entire wing and aircraft shapes. The future is on the wing, and there is no stopping it.

FUEL SAVINGS ARE ON THE WING FOR BOEING 737-700/800s.

Fuel for thought: Every 737 that barrels down the runway with Blended Winglets™ can expect up to 6% better cruise mileage. That's like getting a free tank every 20 trips–depending on routes and loads. To see how much this advanced technology can save your in-service fleet, call (206) 762-1171 or fly to www.aviationpartnersboeing.com. The future is on the wing.™   *Aviation Partners Boeing*
A joint venture of Aviation Partners, Inc. and The Boeing Company

Figure 2.4 Partners advertising barrels of fuel savings

Figure 2.5 Spiroid winglet during test flight

REVOLUTIONARY BENEFITS OF BLENDED WINGLETS

Safety benefits

Increased range, resulting from improved fuel efficiency, provides an added margin of safety to the Blended Winglet-equipped operator. On critical long-haul, over-water flights—such as Los Angeles to Hawaii—a Blended Winglet-enhanced Gulfstream II benefits from a 50% increase in landing fuel reserves due to improved cruise efficiency. Aircraft experts speculate that safety may also be enhanced for aircraft taking off behind Blended Winglet-equipped aircraft because there is less vortex wake turbulence.

Quieter skies

A Blended Winglet-equipped aircraft achieves normal take-off at reduced power settings and will climb faster with a significantly reduced noise footprint. For example, Blended Winglets on a Boeing 737-700/800 allow 39 additional flights per day at noise-restrictive John Wayne Airport in San Diego. Blended Winglet Systems generally reduce the noise-affected take-off area by 7.5%, a particularly important benefit in Europe where take-off and landing fees are based on the noise-affect area.

Fuel efficiency

API Blended Winglets reduce aerodynamic drag by up to 7.5%, which translates into higher fuel efficiency and lower greenhouse gas emissions—benefits *continued* ➜

Fuel savings of commercial aircraft equipped with Blended Winglets				
Aircraft: Boeing	Average flight (hours per year)	Annual fuel per aircraft (US gallons)	Number of operating aircraft	Total annual fuel savings (million US gallons)
737-700/800 New Generation	2,800	55,000	1,045	57,475
737-300/400	2,800	60,000	1,548	92,880
747-200/300	2,000	275,000	318	87,450
747-400	4,750	530,000	582	308,460
757-200	3,000	90,000	976	87,840
767-300	4,500	210,000	605	127,050

that are sustained over the lifetime of the airframe.[*] Long-range cruise fuel efficiency is increased by up to 7%, CO_2 emissions are reduced proportional to fuel savings, and nitrogen oxide (NO_x) emissions are reduced proportionally greater than fuel savings under typical flight conditions. The financial and environmental benefits of winglets depend on the airplane model and configuration as well as the route. Even a relatively small increase in flight range can save substantial amounts of fuel and emissions if a refueling stop can be avoided.

Blended Winglet-equipped airliners, flying a typical 2,000–3,000 hours per year, will save astonishing amounts of fuel (see Table above).

Style, prestige, and performance benefits

"Gulfstream II charter customers feel like they are on a much newer aircraft, so it pays back that way as well," says Air Group President Jon Winthrop.[†] The improved performance of a Blended Winglet-equipped aircraft translates into increased profits for airlines by improving payload and range, decreasing fuel use, and lowering engine-maintenance costs. Range issues are particularly important for aircraft operating from hot or high-altitude airfields, and for long routes where headwinds decrease range. For example, Blended Winglet systems allow South African Airways to add 11–21 more passengers on long-haul flights during hot-day take-off from its Johannesburg hub. At current fuel prices, fuel savings alone pay back the cost of Blended Winglets in two years of commercial use. Additional benefits include routing flexibility, aircraft resale value, and the enhanced image of a stylish, aesthetically pleasing, and environmentally friendly aircraft.

Competitive sales benefits

Blended Winglet technology translates into increased airliner sales for Boeing in highly competitive market arenas where the added performance of this technology is

continued ➡

[*] The winglet will provide sustained environmental performance over the entire life of the airframe. Competing investments in fuel efficiency from engine upgrades, on the other hand, achieve their highest performance when they are new—and their performance degrades as parts wear until the next maintenance cycle.

[†] Quoted in "The Ultimate Winglet," *World Aircraft Sales*, May 1999.

often the deciding factor. One such sale occurred in 2001 when Qantas purchased 15 Blended Winglet-equipped Boeing 737-800s with options for 40 additional aircraft. South African Airways recently ordered 18 Blended Winglet-equipped 737-800s, rather than Airbus 320s, because Blended Winglets provide the range needed to serve important markets non-stop. "People close to the deal said the Blended Winglets offered on the Boeing plane gave it an important performance edge over the Airbus A320 on new long-haul domestic routes planned by Qantas," said one analyst.[‡]

Payload benefits

The 737-800 can carry an additional 6,000 lb (2,720 kg) payload or add 130 nautical miles (240 km) to range. This advantage, in turn, brings in additional revenues, reduces costs, reduces pollution, and has other benefits.

Top-quality materials and construction

Blended Winglets are designed to be as light as possible so that benefits of lower aerodynamic drag can be maximized. API's Blended Winglets are built with carbon-fiber composite material combined with a Nomex® honeycomb stabilizing core, and they use solid laminate spars with aluminum leading and trailing edges.

‡ James Wallace,"Aerospace Notebook: Partnership with Boeing 'Starting to Take Off,' " *Seattle Post-Intelligencer*, February 25, 2002.

Aviation Partners time-line

1492 ● Leonardo da Vinci (Italy) describes a flying machine.

1783 ● Joseph Montgolfier and Étienne Montgolfier (France) launch the first hot-air balloons.

1891 ● Otto Lilienthal (Germany) succeeds with the first reproducible gliding flights.

1897 ● F.W. Lanchester (USA) patents simple wing-end plates for gliders and in anticipation of powered flight.

1900 ● Ferdinand von Zeppelin (Germany) builds the first successful dirigible.

1903 ● Orville and Wilbur Wright (USA) are the first to fly powered aircraft.

1927 ● Charles A. Lindbergh (USA) completes first solo non-stop transatlantic flight from New York to Paris.

1937 ● First experimental pressurized-cabin airplane, a Lockheed XC-35, made first flight at Wright Field.

1947 ● Charles E. Yeager (USA) completes the first supersonic flight in a rocket-powered Bell XS-1.

1949 ● First non-stop, round-the-world flight.

1976 ● NASA's Richard T. Whitcomb invented vertical wingtip extensions as a means to increase lift-to-drag performance.

1981 ● McDonnell Douglas installs winglets on the MD-11 and C-17 Globemaster III.

1983 ● Learjet installs winglets on its business jet.

1988 ● Airbus is first to install winglets on a commercial passenger aircraft (Airbus A320).

1991 ● Joe Clark and Dennis Washington form Aviation Partners (AP) and hire Dr. L.B. Gratzer and a dream team of retired Boeing and Lockheed aerospace engineers.

1993 ● AP installs high-aspect-ratio Blended Winglets on Gulfstream II and immediately markets the retrofit after FAA approval.

● AP flight-tests Spiroid wingtips on a Gulfstream II and achieves more than 10% greater fuel efficiency at cruise but delays commercialization in favor of Blended Winglets that are more stylish and more easily satisfy FAA and aircraft authorities.

● Bernie Gratzer receives patent #5,348,253 for Blended Winglets.

1995 ● Clay Lucy knocks two hours off the Los Angeles to Paris World Speed Record on a Gulfstream II with AP winglets.

1996 ● Clay Lucy establishes seven new time-to-climb records, including a dramatic climb from sea level to 40,000 ft (12,000 m) in just 6 minutes and 20 seconds on a Gulfstream II with AP winglets.

1999 ● AP Blended Winglets tested on Boeing business jet.

● Aviation Partners Boeing (APB) joint venture formed to apply winglets to all suitable Boeing aircraft. The winglet is offered as a standard feature of the Boeing business jets.

● Qantas purchases 15 winglet-equipped Boeing 737-800s and takes options for at least 40 more—citing the winglet as a deciding factor.

2000 ● Hapag-Lloyd is first airline to operate a Boeing 737-800 with retrofit Blended Winglet technology.

● South African Airlines is first to order winglet-equipped Boeing 737-800s, stating that the winglets make the 737 outperform the rival Airbus A320 on long flights.

● Air-Berlin first to fly a Boeing 737 with factory-option Blended Winglets.

2001 ● Blended winglet program announced for Raytheon Hawker 800 business jet with 7% increase in cruise fuel efficiency and significant improvements in take-off, initial cruise altitude, and range.

3

DaimlerChrysler

The champagne of natural refrigerants[*]

Vehicle air conditioning with natural carbon dioxide refrigerant can protect the climate system with higher fuel efficiency and negligible refrigerant greenhouse gas emissions, and it will support the commercial success of fuel cell, hybrid, and direct-injection vehicles where supplementary heating and air conditioning is needed.

DaimlerChrysler's ambition to be the first major automobile manufacturer to have a CO_2 vehicle air conditioner installed on a production vehicle sparked a worldwide quest for a better vehicle air-conditioning system. This quest has created a healthy competition between enhancing the existing hydrofluorocarbon (HFC)-134a systems, choosing a new HFC-152a refrigerant with higher environmental performance, or taking the revolutionary leap to CO_2.

Although CO_2 is a plentiful greenhouse gas that changes the climate, this use as a refrigerant will only constitute a minor emission—and will actually benefit climate protection because it has the potential to improve fuel efficiency dramatically. Furthermore, advocates of CO_2 air conditioning have captured the attention of European and Californian environmental regulators who are demanding prompt action to fight global warming. Carbon dioxide air conditioners can also function as heat pumps to warm and defog new fuel cell, hybrid, electric, and direct-injection vehicles that operate so efficiently that adequate waste heat from the engine is not available to heat the passenger compartment.

* The authors are grateful for interviews and supplementary assistance by the following DaimlerChrysler engineers and managers: Mark E. Dawson, Roland Caesar, and Jürgen Wertenbach; to Obrist Engineering experts Martin Graz, Frank Obrist, and Frank Wolf; to General Motors engineer Bill Hill; to Suntest Engineering President Ward Atkinson; and to EPA's Drusilla Hufford. DuPont's Mack McFarland provided valuable insight and edits.

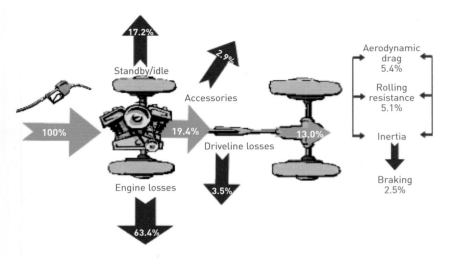

Figure 3.1 How vehicles use and waste energy

The price of "cool"

Air conditioning, or A/C, is considered essential for many transport vehicles, from passenger cars to buses, trucks, and even helicopters. But cool comfort often comes at a high price to the environment, in terms of both ozone destruction and global warming.

From the 1950s until circa 1994, vehicle A/C systems used chlorofluorocarbon (CFC)-12, a potent ozone-depleting substance, as their refrigerant. "Air-conditioner engineers were spoiled by the simplicity of CFCs, refrigerants that are perfect in every way—except they lead to the destruction of all life on Earth," says Frank Obrist, chief engineer of the Wankel Research Institute, an early partner in the CO_2 air-conditioning project.

In response to the Montreal Protocol on Substances that Deplete the Ozone Layer, all new vehicles with A/C sold in developed countries after 1994 have used HFC-134a, a refrigerant that doesn't deplete the ozone layer. Using HFC as a refrigerant has other benefits as well. Its contribution to global warming is six times less than CFC-12; it is non-flammable; and it has low toxicity. Its cooling capacity and energy efficiency can be made comparable to CFC-12 through improved engineering.

But even the current A/C systems that use HFCs continue to contribute to global warming, in four different ways: (1) directly as a result of emission of refrigerant to the atmosphere (e.g. from system leakage, servicing, and vehicle disposal); (2) indirectly from the release of CO_2 from burning fuel to power the A/C system; (3) from fuel consumed to carry its weight; and (4) from the production and recycling of the materials composing the system. The combined effect of these impacts is called the life-cycle climate performance (LCCP) (see Figure 3.2). Overall, the transport sector is responsible for over 25% of all greenhouse gas emissions, with road vehicles responsible for about 77% of that.

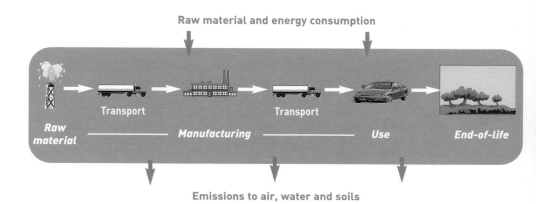

Raw material and energy consumption

Transport

Transport

Raw material

Manufacturing

Use

End-of-life

Emissions to air, water and soils

Figure 3.2 Life-cycle climate performance of mobile air-conditioning systems

HFC-134a air conditioning had matured in the market without incentives for fuel efficiency and leak-tightness. Environmental authorities worldwide ignored the air conditioner in setting standards for fuel efficiency—allowing vehicle manufacturers to satisfy cooling comfort with low-cost systems that use unnecessarily large amounts of fuel.

Although less obvious, the most important contribution that CO_2 air conditioning will make to climate protection is that it empowers the transition to fuel cell, hybrid, electric, and direct-injection vehicles. Energy-efficient cooling and heating are accelerating the market penetration of these new technologies—yielding 55% improved efficiency with fuel cell vehicles, 50% improved efficiency with hybrid vehicles, 25% improved efficiency with electric vehicles, and up to 20% improvement from direct injection. The current HFC-134a A/C systems are unsuitable for these new vehicles because the air conditioning uses too much power, and because the refrigerant greenhouse gas emissions offset part of the gain in fuel efficiency.

Engineering irony: using CO_2 to improve A/C and protect the climate

The challenge faced by the vehicle A/C industry was how to eliminate HFC emissions cost-effectively and at the same time increase fuel efficiency, without adversely affecting cooling performance or reliability. Engineers at DaimlerChrysler decided that the solution might be CO_2—the most prevalent greenhouse gas—but one that could also function as a refrigerant to replace HFCs. If they were right, and if they could overcome the engineering, economic, and regulatory hurdles and win acceptance from their customers, the engineers could foresee dramatic improvements in fuel efficiency. It might have been a gamble, but the high pay-off in climate protection made it worth pursuing.

It is *counter-intuitive* that CO_2 greenhouse gases can protect the climate. Here's how this is possible. The CO_2 used as refrigerant is captured from manufacturing sources such as fertilizer plants and breweries that would otherwise vent to the atmosphere. (Hence, use of refrigerant delays but does not increase the emissions.)[1] Over the life of a vehicle, only one or two kilograms (2–4.5 lb) of CO_2 refrigerant will be emitted. Carbon dioxide systems are expected to be as much as 25% or even 30% more efficient than current air conditioners. And CO_2 air conditioners operated in the heat pump mode to warm vehicles are expected to be far more energy-efficient than electric or combustion furnaces that would otherwise be necessary to heat electric, fuel cell, hybrid, and direct-injection engines that do not produce enough waste heat for these functions. Higher efficiency of heating and cooling reduces the emissions from fuel consumption necessary to power the air conditioner and the furnace.

DaimlerChrysler's quest

Within DaimlerChrysler the quest for CO_2 technology was pursued under the joint support of Roland Caesar and Jürgen Wertenbach, Managers at DaimlerChrysler's Mercedes-Benz Stuttgart research center. Carbon dioxide was first conceived as a refrigerant in 1850, but traditional engineering calculations had failed to recognize the energy efficiency potential of CO_2 systems because calculations were based on unoptimized system components with configurations similar to CFC systems.

It was not until the late 1980s that CO_2 was seriously reinvestigated, after the world began to phase out CFCs. In 1989, Gustav Lorentzen revived the work he and colleagues had done years earlier on natural refrigerants and invented a transcritical CO_2 refrigerant cycle capable of higher energy efficiency. But Gustav Lorentzen's invention came too late to allow the shift from CFC to move directly to CO_2. Wertenbach had his curiosity piqued by lectures by Gustav Lorentzen and Jostein Pettersen explaining the new CO_2 invention and crediting the work of earlier engineers. He acted on his curiosity by assembling a large collection of historic technical papers mostly published in German between 1910 and 1930. He studied

---

1   For example, the "Kellogg" process uses natural gas and air to produce CO_2 and ammonia (NH_3) as feedstocks for caprolactam urea or ammonium sulfate fertilizer, with excess CO_2 vented if side-stream markets are not available. A very small amount of energy is required to compress CO_2 scavenged from manufacturing facilities. The processing and handling emissions are far smaller than those to manufacture other refrigerants.

the collection intently while recovering from a shoulder injury at home—coming to the conclusion that modern engineering, materials and controls could overcome problems and realize the theoretical potential of the CO_2 cycle.

Like others in the vehicle A/C business, Caesar and Wertenbach's initial goal in 1994 was to find a technology to eliminate HFC refrigerants without adversely affecting air-conditioning performance or fuel efficiency. Lacking sufficient funding from their internal budgets, they looked outside and found strong support from the European companies who had worked together on eliminating CFC-12 from vehicle A/C systems. The European Community Industrial and Materials Technology Program funded the new consortium, called "Refrigeration and Automotive Climate under Environmental Aspects" (RACE). The consortium quickly embraced CO_2 as the most technically and economically feasible option and went right to work.

RACE was conceived as a methodical three-year project to explore theory and computer model performance, to develop key system components and optimal layout, and to bench-test prototypes. The initial results were so promising that the team quickly shifted its attention to building two working prototypes: a BMW 520i and a Volkswagen Polo (compact class).

In just three years, RACE saw some extraordinary results. Cooling performance of its prototypes was comparable to the existing HFC-134a systems in highway driving (less comparable in city driving), with the potential to reduce direct refrigerant and indirect fuel greenhouse gas emissions dramatically. It was also determined that the expected higher production costs of this new technology might

RACE PARTICIPATING VEHICLE MANUFACTURERS AND THEIR CONTRIBUTIONS

- BMW (hoses, expansion device)
- Daimler-Benz (computer modeling and system control)
- Rover Group (safety)
- Volkswagen (component integration)
- Volvo (monitoring and testing)

be offset by reduced fuel consumption, lower-cost refrigerants, and simplified service.

But the results of the RACE project could only be realized with the help of separate development by Daimler-Benz of a suitable compressor for the CO_2 refrigerant loop. Fortunately, the encouraging results persuaded Audi, BMW, Daimler-Benz, and Volkswagen to pursue development and continue to cooperate. Perhaps more important, the results had proven to RACE management that each company working on the team got far more than they paid for—both in CO_2 systems engineering and in basic air-conditioning know-how.

This was also made possible because, as Wertenbach emphasizes, his Research Manager Roland Caesar was one of the few business managers who understood not only air-conditioning

Roland Caesar

**RACE PARTICIPATING A/C SYSTEM
SUPPLIERS AND THEIR
CONTRIBUTIONS**

- Behr (heat exchanger, evaporator)
- Danfoss (compressor)
- Valeo Klimasystem—Rodach
- Valeo Thermique Habitacle—La Verriere (condenser and evaporator)

technology but also corporate decision-making, environmental vision, *and* the politics of environmental protection. Caesar is equally generous in his praise of Wertenbach, whom he describes as "a determined and gifted engineer who will never take no for an answer." Both also credit the role played by the Mercedes-Benz racing culture, and their sophisticated customers who appreciate technical elegance.

Carbon dioxide refrigerant in automotive air-conditioning systems[2]

Air conditioners are designed to take advantage of the physical and thermal properties of whatever refrigerant is selected. The advantage of carbon dioxide results from the law of physics that, at any ambient temperature, high-temperature objects cool faster, and more efficiently, than cooler objects. The heat exchanger used with carbon dioxide can operate at a very high temperature because the carbon dioxide is kept as a high-pressure "transcritical" gas while the HFC-134a must be kept at its condensation temperature, which is much lower.[3] The carbon dioxide system has the advantage that a lower volumetric flow is required to achieve a given cooling capacity and that compressing a gas to a higher pressure requires a lower volumetric flow to achieve a given compressor capacity. Transport properties of carbon dioxide allow a reduction of pressure drop for all passive components such as refrigerant lines and heat exchangers. High miscibility of carbon dioxide and lubrication oil allows reduced oil circulation rates in the loop, which contributes to the overall technical advantage of higher cooling capacity and lower fuel use.

---

2   Edited and translated with the assistance and permission of the authors. See Peter Kuhn (Obrist Engineering), Martin Graz (Obrist Engineering), Frank Obrist (Obrist Engineering), Willi Parsch (LuK Fahrzeug Hydraulik), and Frank Rinne (Sanden), "Kohlendioxid-R744 als Kältemittel in Fahrzeug-Klimaanlagen" [Carbon Dioxide R744 as a Refrigerant in Automotive Air-Conditioning Systems], *Automobiltechnische Zeitschrift* (*ATZ*) 103.12 (2001).

3   The heat exchanger, which dissipates the heat, is referred to as a "condenser" in HFC-134a systems but as a "gas cooler" in carbon dioxide systems.

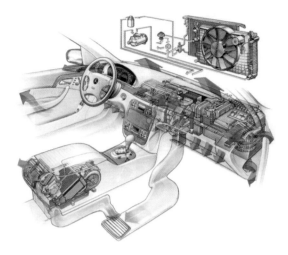

Figure 3.3 Carbon dioxide vehicle A/C system components

Source: Obrist, LuK, Sanden

The challenges of combining science and practical engineering

DaimlerChrysler research proved that higher overall energy efficiency is possible in vehicle applications, and that the benefits are achieved by the thermodynamic and thermophysical properties. They also determined that the weight of the system and the energy used in the manufacture and recycling of components is comparable to current models.[4]

The transcritical CO_2 system also looks similar to today's HFC system, but operates at pressures five to eight times higher than HFC systems.[5] Therefore the introduction of CO_2 climate-control systems requires careful engineering to accommodate high refrigerant pressures and containment and ventilation features to assure that possible leaks of CO_2 refrigerant do not exceed safe concentrations in passenger compartments. Additional components and controls will allow operation at, or near, optimum energy efficiency.

Whereas commercial success requires reliable systems with high cooling performance and reliability, environmental success requires cost-effective reduction of

4   US Patent number 5,245,836 was granted to Gustav Lorentzen, Jostein Pettersen, and Roar R. Bang on September 21, 1993: "Method and Device for High-Side Pressure Regulation in Transcritical Vapor Compression Cycle." Other methods, already in the public domain or patented by DaimlerChrysler and others, may achieve the same performance.

5   Up to five times higher than HFC-134a systems on the high-pressure side of the system and up to ten times higher on the low-pressure side.

greenhouse gases. The DaimlerChrysler system improves fuel efficiency at significantly lower cost than equivalent additional investment in power trains, lightweight materials, or aerodynamics. A cooling system at full economy-of-scale is estimated to cost about $40–180 more than current systems that only air-condition.[6] But transcritical CO_2 systems functioning as a heat pump can also eliminate the need for fuel-fired or electric heaters on fuel cell, hybrid, electric and direct-injection vehicles that do not otherwise have adequate heating in passenger compartments. Furthermore, CO_2 heat pumps integrated into drive trains can warm fuel injectors, catalytic converters, and transmissions rapidly. Warm fuel injectors produce less air pollution; warm catalytic converters reduce tailpipe emissions; and warm transmission oil reduces friction and increases vehicle mileage. Win, win, and win!

Now that it's engineered, how do we sell it to the world?

Experts knew that the transition to CO_2 would be far more complicated than the transition from CFC-12 air conditioning to HFC-134a air conditioning. CO_2 technology requires a highly complex new design and integration by experts in thermodynamics, physics, materials, controls, and human physiology. This transition is made more complicated because climate control is no longer a core business for any of the major vehicle manufacturers.[7] Under these circumstances, it is absolutely necessary to recruit the support of suppliers and academic scientists. But, without some core competence on the manufacturer side, this level of effort would not be offered by suppliers. Fortunately, the suppliers knew that the extraordinary results of the RACE program made cooperation worthwhile.

That persuaded Audi, BMW, Daimler-Benz, and Volkswagen to pursue development, and to continue to cooperate. Perhaps more important, the results had proven to Caesar and Wertenbach and the other RACE partners the need to begin to promote their results and development plans at international conferences, at trade fairs, and in technical publications. Some environmental authorities and non-governmental organizations (NGOs) were supportive, but HFC manufacturers and allied business interests were outraged and threatened. "Experts from the HFC

6   Estimates of additional vehicle manufacturing costs were developed by consensus and presented by Stephen O. Andersen at the *Mobile Air Conditioning Summit: Options to Reduce Greenhouse Gas Emissions due to Mobile Air Conditioning*, Brussels, February 10–11, 2003; and at the *Verband der Automobilindustrie (VDA) Alternate Refrigerant Winter Meeting Automotive Air Conditioning and Heat Pump Systems Conference*, Saalfelden, Austria, February 13–14, 2003. These costs do not include component retooling costs and additional profits that could be charged for these premium A/C systems.

7   In the late 1920s, Thomas Midgley and his associates invented CFCs while working in the General Motors Laboratory for the Frigidaire Division, which manufactured refrigerators.

lobby told anyone who would listen that the RACE companies were technically confused," says DaimlerChrysler. "They presented theoretical explanations that CO_2 air-conditioning systems couldn't be energy-efficient, couldn't provide adequate cooling, and couldn't be repaired safely. They warned us that our mistaken results could also stimulate unnecessary regulation."

In 1997, RACE presented its results at the Earth Technology Forum in Washington, DC. This turned out to be a very fortunate meeting. US EPA managers encouraged the partnership to proceed, and Ward Atkinson, Chair of the Society of Automotive Engineers (SAE) Interior Climate Control Committee, and Simon Oulouhojian, Chairman of the Mobile Air Conditioning Society Worldwide, offered assistance. "Ward and Simon had been the drivers behind CFC recycling and the introduction of HFCs, and they completely grasped the environmental importance of what we were proposing," says Wertenbach.

"Ward is a natural leader, deal-maker, and project juggler; fatherly, honest, influential; and he fears nothing. Ward had the practical experience of vehicle engineering, road-testing, and SAE standards, and Simon was completely open in addressing the challenges of safely repairing high-pressure air-conditioning systems with CO_2 refrigerants." Oulouhojian and Atkinson were among the most globally connected and respected vehicle air conditioning experts who they knew.

Over dinner during the SAE Congress in 1998, Atkinson and his colleagues proposed a "drive test" of the prototypes for July in Phoenix. Phoenix, Arizona, and Cape Town, South Africa, are the two principal cities where air-conditioned cars are tested. "If you can cool them there, you can cool them anywhere."

At the July 1998 meeting in Phoenix, some skeptics became supporters and key air-conditioner suppliers joined the research effort. "Some experts expressed private doubts about the technology, but no company wanted to offend German vehicle manufacturers," says Atkinson. "We were excited by the engineering challenge and the benefits of pushing technical limits." By the end of the meeting, it was decided to meet in Phoenix the next summer. Thus, the RACE member companies transformed the European consortium into the global "Phoenix Forum." Figure 3.4 shows a Phoenix Forum test engineer measuring solar energy.

Next: how to get the "new team" to jump over all the hurdles

Not all the hurdles were technical. "One of the most difficult challenges was to maintain an international team composed of fierce competitors from entirely different national and corporate cultures, and for engineers to learn how to work effectively with managers and environmental regulators," says Bill Hill of General Motors. "No one was a better diplomat and human integrator than BMW's Linda Gronlund. She taught us how to find the right people to say yes without going through all of the pain of transforming corporate culture, and when to proceed on the assumption that silence is agreement."

Figure 3.4 Phoenix Forum test engineer measuring solar energy

Gronlund functioned far beyond her official position as top environmental manager for BMW. Earlier, when she worked for Volvo, Linda was one of the first industry advocates of global cooperation on CFC-12 recycling from vehicle air conditioning. In the early 1990s, she persuaded Volvo to announce a CFC-12 phase-out, earning Volvo the 1993 US EPA Stratospheric Ozone Protection Award. "She was a hands-on advocate of environmental protection," says Atkinson: "A person who knew how to get things done." Roland Caesar adds, "Linda taught us how to understand North American engineers—what to say, what not to say, and when to listen. I really don't know how we will replace her."

Very sadly, Linda Gronlund was killed on September 11, 2001 aboard the United Airlines Flight 93 that was hijacked by terrorists and crashed in Pennsylvania.

One of the first big challenges was that European environmental regulators and customers were demanding improvements in fuel efficiency and fuel efficiency standards that had been constant since 1990. In 1999, experts from environmental ministries were invited to Phoenix, but only the US EPA attended. "EPA has great respect for Ward Atkinson and Simon Oulouhojian," says EPA executive Drusilla Hufford. "They had earned EPA's highest awards for stratospheric ozone protection and they are masters at building international cooperation." Over 100 experts at Phoenix 1999 decided unanimously to asked SAE, the Mobile Air Conditioning Society (MACS), and EPA to formalize a "Mobile Air Conditioning Climate Protection Partnership"[8] and to prepare necessary standards and regulations to encourage the development of new refrigerants, including CO_2.[9]

DaimlerChrysler's strategy of creating a global team was working—the consortium now included US EPA, SAE, MACS, vehicle manufacturers, technical institutes, and automotive suppliers from Europe, Japan, and North America. By 2001 the partnership also included the European Commission and the Environmental Agency of Japan.

How Daimler finally developed the system

"The global consortium was absolutely essential for success," says Caesar. "But the consortium alone could never persuade our management to embrace this new technology." Caesar and Wertenbach built support from the ground up. Managers in every division were briefed and had their questions answered. The extensive briefings and discussions highlighted the challenge of containing the high pressure and lubricating moving parts. Coincidentally, Mercedes had purchased the Technical Institute of Engineering Study (TES) Research Institute (Wankel Institute), created by Felix Wankel, and was looking for worthy projects. Caesar and Wertenbach seized the opportunity.

Felix Wankel was a largely self-taught engineer who relentlessly pursued the development of the rotary engine, from his initial ideas in 1926, to the first working model and the founding of his company in 1957, through extensive licensing, and to its most successful commercialization in the Mazda RX7.[10] In 1973, Mercedes-Benz, under the leadership of Dr. Hans Liebold, was about to release the entirely new MB-CIII sports car featuring a Wankel engine, but abandoned the

8   Organizing partners are: SAE, US EPA, and MACS Worldwide. Government, academic, environmental, testing, and association partners are: Automotive Aftermarket Industry Association, Ecole des Mines de Paris, Environment Canada, European Commission, International Organization for Standardization, Ministry of Environment Japan, Underwriters Laboratories, University of Braunschweig (Germany), University of Illinois (USA), University of Maryland (USA), US Army, US National Renewable Energy Laboratory, and World Resources Institute. Corporate partners are: AC Delco, Audi, Behr, Bergstrom, BMW, CalsonicKansei, Clore, DaimlerChrysler, Delphi Corporation, Denso, DuPont Fluoroproducts, Eaton, Fiat, Four Seasons, Freightliner, General Motors, Goodyear, Honda, Johnson Controls, Kia, Konvekta, Mitsubishi, Modine, Nissan, Neutronics, Parker-Hannifin, Sanden, Snap-On Diagnostics, SPX Robinair, Subaru, Texas Instruments, Toyota, Tracer Products, UView Ultraviolet Systems, Valeo, Visteon, Volkswagen, and Volvo Car Corporation.
9   Another option is to use a flammable refrigerant, like propane or HFC-152a, that has a low global-warming potential. However, these refrigerants are not suitable in heat-pump mode.
10 One measure of the potential of the Wankel engine is that it was licensed by so many companies: Alfa Romeo, American Motors, BMW, Curtis-Wright, Ford Motor Company, General Motors, Johnson Outboards, Kawasaki, Mazda, Mercedes-Benz, Nissan, NSU, Ingersoll-Rand, Outboard Marine, Perkins, Porsche, Rolls Royce, Suzuki, Toyota, Volkswagen, Toyo, Yamaha, and Yanmar, to name a few.

MOBILE AIR CONDITIONING CLIMATE PROTECTION PARTNERSHIP

SAE in cooperation with the International Organization for Standardization (ISO) organized the Alternate Refrigerants Task Force of the Interior Climate Control Standards Committee to write standards necessary for commercialization of new refrigerants and to sponsor the Phoenix Forum. The SAE Cooperative Research Program (CRP) is benchmarking the energy performance of HFC-134a systems and system testing experimental new CO_2 systems.

Figure 3.5 Wankel engine combustion chamber

project when the oil embargo made fuel economy more important than power. The Wankel engine had long captured the imagination of engineers and scientists because of its elegant mechanism, theoretical simplicity, and the nearly impossible problems of machining and lubricating parts to seal the combustion chamber (see Figure 3.5).[11]

The Wankel research institute's chief engineer, Frank Obrist, assembled a team of his best engineers. They quickly mastered the physics of refrigeration and applied the lessons of the Wankel to the compressor design. When the Wankel Institute closed in 1998, Obrist formed his own consulting company and continued his work with DaimlerChrysler to develop a new drive mechanism for an A/C compressor offering high efficiency and excellent control behavior.

After the initial development with DaimlerChrysler, Obrist worked with Audi and supplied its CO_2 compressor to the US Army.[12] Obrist eventually supported Sanden, a Japanese A/C compressor company, and LuK, a German hydraulic pump company, to produce a durable compressor that is lighter and smaller than the same-capacity HFC-134a compressor. Smaller compressors and smaller refrigerant hose diameters have "packaging" advantages for fitting the air-conditioner systems into the increasingly small space of engine compartments.

"At Wankel we had the 'rocket science' that made CO_2 air conditioning possible," says Frank Obrist. His colleague Caesar responds, "Frank perfectly complemented the Mercedes-Benz research center engineers; Jürgen knew how to integrate the system in the vehicle and Frank could make the components work."

11   The Wankel engine has only three main moving parts (two rotors and the eccentric shaft) that rotate continuously in a single direction, surrounded by the peripheral housing where the rotor tip traces out an epitroichoid curve. See www.monito.com/wankel/rce.html.

12   The US Army (USA CECOM) is developing CO_2 systems for stationary, wheeled vehicles, and individual soldiers. A full-scale high-mobility multipurpose wheeled vehicle (HMMWV) prototype is planned for completion sometime in 2003. See www.cartech.doe.gov/research/systems/air-conditioning.html.

In the early 2000s, the S-Class Mercedes appeared to offer the perfect opportunity to introduce the new CO_2 technology. The cooling and heating performance of the CO_2 technology had been proven, Daimler engineers believed that fuel efficiency would be higher, and the technology was considered a marketing advantage. The S-Class Mercedes-Benz appeared to be the premier platform for technical innovation as well as for marketing those innovations. "Mercedes customers appreciate the finer things in life, including technical elegance and environmental leadership. It is only natural that we place technology from our racing and environmental programs on this high-performance vehicle," the company says.

The S-Class has technical and economic advantages for this undertaking. It already has many of the sophisticated air-conditioning monitors and controls that would be necessary to operate the more sophisticated CO_2 systems to assure comfort and reliability. The fuel savings are more vital to the C-Class and E-Class for meeting the pledge of 25% greater fuel efficiency by 2008. Furthermore, the S-Class is scheduled for major design and technical upgrades around 2006/07. The S-Class has a legacy of environmental pioneering—it was also the first car in the world to shift from CFC to HFC in 1990.

Cutting-edge technology is often introduced on one vehicle and then simplified for application on other models; once the bugs are worked out, controls are simplified, and costs are brought down. This was the strategy for Mercedes' first-in-the-world introduction of anti-lock breaks and airbags, and it will be the strategy for CO_2 air conditioning. "Customers are willing to pay for front and side airbags because they reduce the risk of serious injury; they will also want to pay for climate-protecting CO_2 air conditioning when they realize the importance to their children and grandchildren," the company says.

The standard strategy was to learn from the S-Class, and then quickly apply it to other platforms. The Chrysler PT cruiser is the cross-ocean prototype that is helping engineers in North America contribute to the Mercedes product launch and prepare for the follow-up. After the Mercedes S-Class launch, the technology could be implemented in other models as other vehicle platforms are updated.

The A/C component and system-integration companies that supply Mercedes and other automakers intend to globalize the technology first in markets with high fuel prices and stringent regulations, and in locations where the added comfort and safety of rapid warm-up are most appreciated. With adequate incentives, CO_2 systems could penetrate up to half of the global market by 2010, and could dominate by 2015.

But there were hurdles that delayed the launch of this new technology on the Mercedes S-Class in 2000, including the start-up costs of the new technology, questions of reliability, and slower-than-expected incentives from government regulators. Sometimes it is more courageous and wise to delay commercial introduction of a new technology until the moment is right. There was always the next year. "Surely 2001 would be the year to commit to the launch of the new CO_2 technology," thought the team members.

A new "bump" in the road

By March, 2001, the problems were almost solved and engineers were frantically preparing cars for performance testing. "Unfortunately, the Mercedes research committee had a budget shortfall and chose to eliminate our project from a long list, without even consulting us. We were crushed by the news. Our managers told us that an appeal to the research committee would not be possible, and ordered us to remove the test car from the air-freight manifest for vehicles going to the Arizona test track," says Caesar. When news spread that Mercedes engineers and suppliers would no longer be available to the team, Audi, BMW, and Volkswagen began to shut down their programs as well. All might have been lost. "But," remembers Caesar,

> instead of giving up, we found a strategy to get the cars to Arizona without air transport. At the test track in February and July, and in test labs before, we documented the cooling, heating, and fuel consumption at the test track, and then drove the car to the DaimlerChrysler Board Meeting in Houston. The order was to not appeal the decision to the research committee . . . we followed the letter of that order by bypassing the research committee. Still we worried what would happen.

Mercedes-Benz has a long history of building and retaining customers by taking risks with extraordinary new technology and style. It is also proud of community insubordination. For example, at their museum in Stuttgart, guides usually tell the story of how the young wife of Karl Benz left a note for her sleeping husband explaining that she had taken their two small children to see her mother in his experimental car—a trip distance far exceeding the range that even the most daring men would attempt. Caesar continued:

> We had never been more nervous than when we got to Houston. We asked one very experienced director, Professor Gaus, to take a test ride. He took the car out on his way to morning golf, with some other senior managers in the passenger seats. He liked the powerful cooling and said S customers would love it—but somehow fuel efficiency was on his mind. He kept asking me about fuel use and I kept showing him the test results proving a 25–30% improvement. Then he told us we must proceed with the work. I quickly explained how I had bent the rules. Later on, he influenced the board to support the project, explaining that this technology could be a key element of climate protection.

Caesar's partner, Wertenbach, describes the team's thinking during the development of the system:

> My research team had naturally concentrated first on the technical performance—cooling and heating—but had failed to realize the importance of fuel efficiency. As soon as we started thinking about fuel emissions we were drawn to the opportunities for whole-vehicle energy management. That's when we realized that the heat pump could be used to drastically reduce energy consumption by rapidly heating the transmission to reduce friction, the catalyst to increase chemical cleansing of exhaust, and cold heat or cool the combustion process. We also began to

appreciate that it is more cost-effective to reduce the heating and air conditioning energy consumption of new fuel cell vehicles than it is to add more fuel cell capacity (and weight) to provide comfort and safety.

Environmental benefits also sell cars

The new DaimlerChrysler CO_2 technology is applicable to trucks, buses, cars, and helicopters. It is currently most economical in four-passenger or larger vehicles and light trucks used in locations where both heating and cooling is required. It will be implemented first in vehicles where adequate space is easily available and for designs that attract customers who appreciate technical and environmental innovation. The first introduction will be in prestigious Mercedes cars, which will help present the image of technical superiority and reliability. Recall that Mercedes was first to introduce other environmental and safety technology, including safety seats and door latches, diesel vehicle engines, fuel injection, anti-lock breaking systems, and airbags (see the time-line overleaf).

Because of these remarkable engineering successes and potentials, the marketing division of DaimlerChrysler is now taking a stronger interest in environmental performance. "Today, a woman makes the decision in over half of vehicle purchases, and by every research measure women care greatly about the environment and about children," says the division.

Furthermore, Mercedes has always flourished by anticipating markets and getting there first with superior products. Thanks to Mercedes and RACE, now it's "environmentally cool" to be cool.

In late 2002, the final decision had to be made on whether to introduce CO_2 cooling-only systems on the redesigned S-Class, or whether to wait for a model where the system could both heat and cool. The team knew that timing would continue to be key to a successful introduction, and to their ability to show financial success. Once again, the Daimler-Chrysler decision-makers were faced with a new set of hurdles. First, the A/C systems suppliers bid higher than the company reservation price for the new technology. Then the service team raised questions on how to test the systems for leaks when carbon dioxide is prevalent in the air. And, finally, the European Commission was unable to confirm its intentions either to control HFC-134a or to reward HFC-free systems.

As this book goes to press, the timing of the first DaimlerChrysler introduction has been

Small increases in mobile A/C fuel efficiency can translate into enormous global savings. Typical cars operating in hot and humid American cities consume an average of 6% of fuel for air conditioning alone. If all new cars and light trucks sold in the United States in 2000 came with a CO_2 air conditioner, 20.57 million kg (20,570 metric tons) of HFC-134a would be avoided (equivalent to 16.85 MMt CO_2) and the added efficiency would save 62 million gallons of gas, resulting in a total of 0.55 MMt CO_2 reductions.* The combined emissions avoided would be 17.4 MMt CO_2-equivalent, the amount sequestered by 5.2 million acres (2.1 million hectares) of forest.

* This assumes that air conditioning operates 25% of the time, and CO_2 has a 2% higher fuel efficiency than HFC-134a.

postponed and remains uncertain. But it is clear that the pioneering efforts of the DaimlerChrysler team have already made a significant contribution to climate protection by inspiring engineering genius and creating a market for fuel-efficient air conditioning. On December 4, 2002, Toyota's fuel cell hybrid vehicle (FCHV) was launched for street testing with a Denso electrically driven, hermetically sealed, air conditioner/heat pump using natural carbon dioxide refrigerant—Denso engineers credit the European RACE program and German leadership for its birth. "Denso successfully reduces the impact of vehicles on the environment with the world's first CO_2 air conditioner applicable to commercially produced vehicles," said Denso's Satoshi Watanabe. "The new air conditioner will not only protect the environment but also strengthen our global competitiveness."[13]

In February 2003, executives of the European Commission concluded after two days of technical presentations at the MAC Summit that "HFC-134a is an unsustainable option for mobile air conditioning that could be phased out within the decade."[14] This conclusion reveals the impact that DaimlerChrysler's work has had on the future mobile air conditioning market.

DaimlerChrysler time-line

1850   A. Twining receives a British refrigeration patent that mentions CO_2 refrigerants.

1866   Thaddeus S.C. Lowe makes the first recorded use of CO_2 as a refrigerant.

1883   Karl Benz founds Benz & Cie and builds the first automobile.

1884   W. Whiteley demonstrates the first vehicle air conditioning for horse-drawn carriages using a wheel-driven fan to blow air across ice.

1889   Gottlieb Daimler's first car.

1926   Benz & Cie and Daimler Motoren Gesellschaft merge to be Daimler-Benz.

1928   CFCs invented at General Motors Research Laboratory.

1936   Felix Wankel receives his first patent for the "Drehkolben Maschine" (DKM) rotary combustion engine, which is the precursor of the modern "Kreiskolbenmotor" (KKM) engine.

1939   Packard produces the first vehicle with CFC air conditioning (HCFC-22).

1973   High-performance Wankel rotary engine developed with Daimler-Benz is pulled from production in response to the oil embargo with high gasoline prices and regional shortages.

1974   Molina and Rowland publish hypothesis that CFCs are destroying the ozone layer that protects life on Earth from harmful ultraviolet radiation.

13   See Rick Barrett, "Carbon Dioxide May Be Tomorrow's Refrigerant," *Milwaukee Journal Sentinal*, September 21, 2002 (www.jsonline.com/bym/news/sept02/81643.asp); and Denso press release, December 2, 2002.

14   See http://europe.eu.int/comm/environment/air/mac2003/index.htm.

1985 • Vienna Convention.

1987 • Montreal Protocol on Substances that Deplete the Ozone Layer.

1989 • Gustav Lorentzen files for first modern patent of CO_2 system.

1990 • S-Class Mercedes is first production vehicle with CFC-free air conditioning (General Motors, Nissan, and Volvo also among first).

1992 • Framework Convention on Climate Change signed by countries at the Rio Earth Summit.

1993 • Gustav Lorentzen and colleagues granted patent #5,245,836.

1994 • Global production of vehicles almost CFC-free (with exceptions for some countries).

1994-97 • European Union sponsors RACE consortium to develop and demonstrate CO_2 vehicle air conditioning.

1996 • Shared Mercedes/BMW patent for CO_2 heat pump.

1997 • Kyoto Protocol on Climate Change signed in Kyoto, Japan.

1998 • First Phoenix Forum on alternatives to CFC air conditioning.
• Daimler-Benz and Chrysler merge to become DaimlerChrysler.

1999 • Second Phoenix Forum on alternatives to CFC air conditioning.
• Phoenix Forum decides by consensus to create the "Mobile Air Conditioning Climate Protection Partnership" and appoints Stephen Andersen (US EPA), Ward Atkinson (SAE), and Simon Oulouhojian (MACS Worldwide) as Chairs.

2000 • Third Phoenix Forum on alternatives to CFC air conditioning.

2001 • SAE creates new project on benchmarking CFC vehicle air conditioning and testing of alternatives under its Cooperative Research Program (CRP) with US EPA and Environment Canada the first financial contributors.

2002 • Fourth "Phoenix Forum" on alternatives to CFC air conditioning.
• China stops production of vehicles with CFC air conditioning (01/31/02).

2003 • European Communities Commission executives declare HFC-134 "unsustainable" and urge strong measures to reduce direct and indirect greenhouse gas emissions from vehicle A/C.

4
Energy Star
Money isn't all you're saving*

Using state-of-the-art marketing, this government
branding program promotes energy-saving appliances
with empowered customers, rewarded companies, and the
clout of government procurement.

The Green Lights® and Energy Star® programs of the US Environmental Protection
Agency (EPA) and Department of Energy (DOE) are successfully reducing emissions
of greenhouse gases (GHGs) that contribute to global climate change. These
partnerships capitalize on the nation's creativity. They are transforming markets
by enhancing demand for energy-efficient products and services across all sectors
of the economy, driving investment in energy efficiency, and saving money for
consumers and organizations on energy bills. As noted by Christine Todd Whitman, the Administrator of the EPA,

> Since taking office, President Bush and I have emphasized the need to
> build partnerships across traditional boundaries and encourage coopera-
> tion in our efforts to protect the environment. The results have shown
> that environmental achievement and economic prosperity can go hand
> in hand.[1]

Green Lights and Energy Star were brainchildren of the EPA managers who had
successfully used industry cooperation, voluntary agreements, and corporate
pledges to help protect the ozone layer in earlier years. Green Lights was created as
a program to enlist building owners to invest in new lighting proven to save them
money. Energy Star started with desktop computers and has been continuously
expanding its label to new products every year.

* The authors are grateful for interviews and supplementary assistance by former and
current EPA managers John Hoffman, Kathleen Hogan, Brian Johnson, Caley Johnson,
Stephen Seidel, James Sullivan, and Maria Tikoff Vargas.
1 EPA Climate Protection Partnerships Division, *2000 Annual Report*.

Many small devices make light work

In the late 1980s, people came to love laptop computers because they embodied fast-paced independence, and because they represented the future. If computer engineers could make such a monumental improvement only a few years after customers bought their first Macintosh Classic desktop computers, just imagine what they would do next! But there were problems with the first portable computers— they were heavy and bulky, they needed more memory and better screens, and the battery lasted only minutes, not the hours necessary to get a job done.

The computer, component, and software designers did not take long to come up with integrated solutions. Components were made more energy-efficient, batteries were made long-lasting with exotic metals and assembly, and, most importantly, the microchips and software were improved to allow what we today call "sleep modes." Sleep modes save battery power by shutting down features strategically when they are not in use. Disk drives stop spinning when not accessed; screens dim and turn off when the keyboard is inactive; and on-screen data is automatically stored for prompt recovery when the computer is "awakened" by a keystroke or movement of the mouse. This energy-saving technology was applied expertly to portable computers, but not to desktop computers connected to unlimited power via wall outlets.

John Hoffman, the former Director of the Stratospheric Ozone Protection Division and Founder of the Climate Protection Division at the EPA, believed that the long journey toward stabilizing global climate would begin with small steps. Hoffman's strategy was to choose obvious first steps that would demonstrate how easy the journey could be if people took advantage of the opportunities in front of them. He realized that desktop computers left on day and night across the country were consuming an enormous amount of electricity.

He wanted to find a way to encourage computer manufacturers to incorporate the sleep mode into all computers, copy machines, fax machines, and other office equipment. More importantly, he wanted to demonstrate that repeatable improve-

- If just one room in every home in the USA used Energy Star lighting, the change would keep one trillion lb (0.45 trillion kg) of greenhouse gases out of our atmosphere.
- If half of the households in the USA replace their regular TV with an Energy Star-labeled model, the energy savings will be the equivalent of shutting down an entire power plant.
- If just one US household in ten bought Energy Star heating and cooling equipment, the change would keep more than 17 billion lb (7.65 billion kg) of pollution from our air.
- If just one new US home in ten earned the Energy Star label, the change would be equal to eliminating the pollution generated by 600,000 cars for one year.

John Hoffman

Figure 4.1 Energy Star lighting fixtures and bulbs advertisement

ments to institutional or organizational problems could, in aggregation, eliminate the inefficiency that causes global climate change.

Hoffman knows technology, and he knows how to get things done. After graduating from Massachusetts Institute of Technology (MIT) graduate school in systems analysis, he started a consulting business based on one of the first computer-aided, multi-attribute geographic decision-making models. The model allowed companies to choose successfully between alternatives when many objectives needed to be satisfied, whether for locating shopping malls or helping bus companies decide routes and frequency of trips. In his late twenties, he sold the business and retired to San Francisco, where his future wife, Lucinda, practised law and danced in the ballet to the drums of Richard Feynman.[2]

Retiring before 30 is a spectacular accomplishment for most people, but you would almost expect it from Hoffman. He is savvy about business and people; and ideas and insights pour out of Hoffman at a fantastic rate. After half an hour of friendly conversation, you wonder if he will be sending you a consulting bill. When John and Lucinda inevitably left the relaxed coffeehouses of San Francisco, he decided to spend the next segment of his life in public service at the EPA; and she joined a Washington DC law firm.

Hoffman's first job at the EPA was to explore the use of market mechanisms to replace typical EPA "command and control" regulations that were often costly and burdensome. Next, he managed a climate-change policy team that made the first projections of near-term global warming and sea level rise. Their published reports, *Can We Delay a Greenhouse Warming?* and *Projecting Future Sea Level Rise* (both in 1983), were extremely controversial among scientists who were advising on policy matters, and landed John's work on the front page of the *New York Times*.

To generate energy, the majority of power plants worldwide burn fossil fuels. As they burn, fossil fuels emit pollutants such as nitrogen oxides, sulfur dioxide, and particulate matter. They also emit CO_2, the most abundant greenhouse gas in the atmosphere. Our atmosphere today contains about 660 billion more metric tons of CO_2 than it did before industrial times.

The recent *Third Assessment Report* of the Intergovernmental Panel on Climate Change (IPCC) projects that the ongoing accumulation of greenhouse gases will result in a global temperature increase of between 2.5 and 10°F (1.4–5.6°C) by the end of the century. The range represents uncertainties in the growth of emissions over the next 100 years and the effect on global temperatures of a given increase in atmospheric greenhouse gases. However, even the low end of the projection would be an unprecedented change in temperature relative to the past 10,000 years.

Global surface temperatures have already increased by 1°F, and the IPPC states, "There is new and stronger evidence that most of the warming observed over the last 50 years is attributable to human activities."[*] It is unclear exactly how the impact of global warming will affect us, but there are strong associations between the change in temperature and extreme hot days, less extreme cold days, intense rainfall, sea-level rise, and shrinking ice and glaciers. Human health, agriculture, water resources, coastal areas, ecosystems, and wildlife are vulnerable to these changes. The greenhouse gases released today will remain airborne for decades and centuries.

* *IPCC Third Assessment Report: Climate Change 2001*

---

2   The Nobel Prize-winning physicist, described in the best-selling book, *Surely You're Joking Mr. Feynman!* (New York: W.W. Norton, 1985)

Figure 4.2 Kinko's and EPA promote energy efficiency

At that time, under the duress of "15 minutes of fame" and of headaches from publicity about his reports, Hoffman was thinking about moving to the private sector. But he became re-motivated for public work by the issue of stratospheric ozone depletion. Alan Miller, a public-interest attorney at the Natural Resources Defense Council, asked EPA to regulate CFCs as proposed in an old Advance Notice of Proposed Rulemaking (ANPR). The EPA Toxics Office wanted to dismiss the ANPR, basing their decision on a National Academy of Science (NAS) report that endorsed the chemical industry claim that the ozone layer would be safe because the global market for CFCs would never increase to dangerous levels.

Rejecting the NAS conclusion, Hoffman and a small EPA team persuaded EPA top managers to overrule the decision from the Toxics Office; and Hoffman soon found himself in charge of stratospheric ozone protection. In less than five years in that role, Hoffman helped get CFCs controlled internationally under the Montreal Protocol, with the US regulatory program implemented through a market-based program that welcomed voluntary initiative from industry.[3]

Energy Star

With ozone-layer protection well under way, and with confidence in voluntary programs and market transformation, Hoffman turned his attention back to the challenge of improving energy efficiency to protect against climate change.

Drawing from the field of market psychology, Hoffman recognized that the best way to get manufacturers to put sleep mode into their desktop computers would be to offer a solution that could increase profits and market share. His idea was that the EPA would promote a label to distinguish energy-efficient desktop computers, attracting consumers and giving environmentally superior computers an edge in a fast-growing and increasingly fierce market (Figure 4.3).

Computer buyers were already familiar with simple third-party endorsements of technical performance and reliability for such attributes as computer operating and core memory, disk storage

Figure 4.3 Energy Star hang tag for appliance labeling

3   Stephen Andersen, working under John Hoffman, organized the first voluntary agreements and corporate leadership programs. Durwood Zaelke was project manager of the international "Pathfinder" meetings that devised strategies to halt the use of ozone-depleting substances in the developing countries and the Union of Soviet Socialist Republics. See Chapter 9 on Trane documenting how the CFC phase-out in building air-conditioning chillers encouraged energy-efficiency improvements that often pay the total cost of replacing old CFC equipment with electricity savings.

capacity, chip speed, and measures of computational speed. It was time that buyers learned which computer models were good for the Earth because of improvements to reduce the unnecessary consumption of electricity.

Hoffman probably comes up with a thousand great ideas in a year, and you can tell how warmly he feels about the people who implement them. Cathy Zoi, Miin Liew, and Bryan Johnson were the force behind the first implementation of what came to be known as Energy Star. Zoi and Johnson negotiated with computer manufacturers and cleared the way at EPA; Liew came up with a logo, presentation graphics, and the first government advertising strategy patterned after the best private-sector models. Together, they translated the unending flow of Hoffman's ideas and energy into something that worked. Zoi and Johnson worked so well together that they stayed together as a team after they left the EPA to work for President Clinton's administration.

Zoi's first move for Energy Star was to contact IBM, Apple, and Intel. She met with them separately to "persuade and build consensus," and to keep them from discouraging each other. She found engineers at Intel who wanted to promote the sleep mode, but they hadn't been able to get any of the higher managers interested. She persuaded managers at IBM and Apple that a label promoted by the EPA could provide an environmental and market advantage. Zoi liberated the engineers at Intel and motivated the managers at IBM and Apple.

Just as the EPA was about to roll out Energy Star, the Electronic Industries Association (EIA) threatened to derail the program. The trade association was skeptical that anything good could come from EPA and felt committed to fight the new initiative. Hoffman has never liked the "cold war" that often characterizes decision-making in Washington, joking that it reminds him of a *Star Trek* episode where two aliens are trapped in a tunnel between two universes, pointlessly battling for an eternity. In the end, however, the involvement of the EIA actually strengthened the program by allowing enough time for more companies to participate in Energy Star immediately after its inauguration.

Energy Star was an immediate success, and by the end of its first year it had 50–60% of the US market participating. This success was multiplied when a 1993 government executive order was released that requires all computers, monitors, and printers purchased by federal agencies to be Energy Star-compliant.[4] After that, manufacturers that depended on the enormous federal market jumped on board quickly. By 1999, 95% of monitors, 85% of computers, and 99% of printers sold in America were Energy Star-compliant.[5] Energy Star gained international attention as well when companies marketed their best products worldwide.

Ambitious young managers from some of the best MBA schools staff the EPA Energy Star team. Naturally, they use business strategies to accomplish government objectives. One of the memorable successes was an advertisement directed to CEOs in *Fortune* magazine (Figure 4.4).

Today, marketing people consider it very important to include Energy Star in their product advertisements. They know that the recognized, trusted logo will not

4   Executive Order 12,873: "Federal Acquisition, Recycling and Waste Prevention."
5   C.A. Webber, R.E. Brown, and J.G. Koomey, "Savings Estimates for the Energy Star Voluntary Labeling Program," *Energy Policy* 28 (2000): 1,137-49.

Figure 4.4 Energy Star protecting the bottom line

only help capture their audience's attention, but it will allow consumers to trust the product. Retailers agree: Sears recently had a program to teach its sales force how to use Energy Star to raise sales.

The strategy of partnerships

The EPA's voluntary Energy Star programs, like private-sector start-ups, are based on market surveys and expert advice. Dave Chittick, Vice President of AT&T, was one of John Hoffman's most influential (and free) consultants. Chittick was a founding member and first chair of the Industry Cooperative for Ozone Layer Protection (ICOLP)—the father of EPA voluntary partnerships. ICOLP was a consortium of two dozen global electronics and aerospace companies and their governments that worked together to select and implement technologies to replace ozone-depleting solvents. Together, they eliminated the solvents, commercialized profitable alternatives, and cooperated with developing countries to transform manufacturing. Chittick gave the EPA Energy Star team a bit of tough love. Hoffman explains:

> Dave agreed with their appraisal of the market potential for efficient lighting, and he agreed that companies could profit from the investment, but he was appalled by our first approach to marketing that quantified benefits to climate protection. Dave showed us how this approach could alienate companies that wanted to help the environment but disagreed about whether the climate is changing. He said we needed to do things differently. He helped us focus our pitch on practical business considerations and the generic environmental benefits that would be created by efficiency. Dave helped us hone our core strategy of getting corporations to sign a contract that committed them to testable targets (such as 90% of building space upgraded to efficient lighting within five years). When Dave left EPA, project planners were exhausted—but we followed his advice and the rest is history.

Hoffman's approach to selling efficiency to America was to remove the market barriers that prevented good investments from being made. Using the insight of Dave Chittick and others, the EPA launched a two-pronged strategy of selling to top corporate managers while assuring support from the facility managers who were in charge of lighting. Therefore, when the EPA laid out the economic benefits of efficient lighting during sales meetings, the sales presentations also focused on how impossible it was for facilities managers to get the resources they needed to do the job. One of Hoffman's ingenious solutions for making electricity savings important to companies was to make the facilities engineers more visible and more important to management.

"One way to respect companies is to assign the most talented government experts to be their point-of-contact," says Hoffman. "For example, Maria Tikoff Vargas had demonstrated extraordinary people skills in organizing the international meetings where the Montreal Protocol was planned. Maria knew how to make people from all cultures feel welcome and how to organize an agenda so that we actually got to a consensus before we adjourned a meeting. She was the best, so we assigned her to Green Lights." Vargas describes her work glowingly:

> Working for the EPA's Energy Star program is exactly like being in the marketing department of an ambitious company. EPA groomed me to operate like successful companies, encouraged me in getting my Master's, and plunged me back into sales. I worked to promote Green Lights in California where I went company to company signing up partners and following through on their lighting upgrades. Then EPA brought me back to Washington where I managed several areas before being promoted to Energy Star brand manager. I have had a great career with the ultimate satisfaction of making the world a better place for future generations.

The EPA team created environmental playing cards with a photo of the facilities manager of each company on the front, and how much money was saved and pollution prevented on the back. EPA would write letters to the CEO and congratulate them for the use of Green Lights, and would quantify the savings and environmental benefits. Typically, the CEO would in turn write a letter of recognition to the facilities manager. Mobil Oil, who became a leading advocate of saving energy, even put an op-ed in the paper about their use of Green Lights, praising their facilities manager as a key player.

When Hoffman left the EPA in 1995, Kathleen Hogan became the Division Director.[6] In its first year under her leadership, EPA added fax machines, copiers, residential heating and air-conditioning equipment, thermostats, new homes, and exit signs to the labeling program. A year later, the Department of Energy got on board, and the two government agencies formed a partnership. DOE quickly added refrigerators, room air conditioners, clothes washers, and dishwashers. In 1997, EPA added scanners, multi-function devices, and residential lighting fixtures, followed by TVs and videocassette recorders (VCRs) in 1998. In 1999, EPA further extended the label to cover commercial and industrial buildings.

By 2001, Energy Star had a total of over 35 product categories, including more than 13,000 product models, representing more than 60% of energy use in the average household. These products deliver the same or better performance than comparable models, while using less energy and saving money. Energy Star also provides easy-to-use home and building assessment tools so that homeowners and building managers can start down the path to greater efficiency and cost savings.

Energy Star, by any account, is a highly successful program. To date, more than 750 million products with the Energy Star label have been purchased, and thousands of companies are working with the EPA to adopt more energy-efficient practices. In 2001 alone, Energy Star helped save enough energy to power ten million homes and reduce air pollution equivalent to taking twelve million cars off

---

6 Today, John Hoffman runs a group of small invention-driven companies that have patents on promising technology, including a new method for air-conditioning buildings that will cost less to install and save over 50% energy; a technology for enhancing the output of the most efficient and environmentally friendly gas-fired power plants; and a solar air conditioner.

EPA's ENERGY STAR® *Transformer Program is a voluntary effort that helps utility companies reduce air pollution and protect the environment. By choosing to purchase transformers with the* ENERGY STAR *label, utility companies can take an active part in saving the earth and reduce emissions by 10 to 40 percent. The program is an effective way for utilities to ensure economic efficiency, protect the environment, and lower customers' monthly bills. For more information, call the toll-free* ENERGY STAR *Hotline at 1-888-STAR-YES (1-888-782-7937).*

Figure 4.5 Energy Star Transformer ad for use by electric utilities

the road—all this for the environment while saving Americans $6 billion on their energy bills and without sacrificing product features, quality, or personal comfort.[7]

"Cooperation from a wide variety of partners has helped make Energy Star the leading symbol of energy efficiency around the world, and a model for partnership programs in the future," says Christine Todd Whitman.[8] "With an annual energy bill of nearly $500 million, Verizon knows that saving energy means saving money. Energy Star is helping us learn where and how our energy investments can reap the greatest savings, which is good for our bottom line and good for the Environment," says Ivan Seidenberg,[9] President and CEO of Verizon Communications.

YOUR NEXT NEW HOME CAN BE
A CHANGE FOR THE BETTER

ASK ABOUT
ENERGY STAR

EPA VOLUNTARY PARTNERSHIPS

- Agricultural partnerships
- Climate leaders
- Coalbed Methane Outreach
- Combined Heat and Power Partnership
- Green Power Partnership
- HFC-23 Emissions Reduction Program
- Landfill Methane Outreach Program
- Mobile Air Conditioning Climate Protection Partnership
- Natural Gas STAR Program
- PFC Emission Reduction Partnership for the Semiconductor Industry
- SF_6 Emission Reduction Partnership for Electric Power Systems
- SF_6 Emission Reduction Partnership for the Magnesium Industry
- Voluntary Aluminum Industrial Partnership

Figure 4.6 Energy Star homes: Las Vegas advertisement

---

7 For a description of sophisticated techniques for calculating financial and environmental benefits of Energy Star, see: C.A. Webber, R.E. Brown, and J.G. Koomey, "Savings Estimates for the Energy Star® Voluntary Labeling Program," *Energy Policy* 28 (2000): 1,137-49. For an example of earlier analysis, see B. Nordman, M.A. Piette, B. Pon, and K. Kinney, *It's Midnight . . . Is Your Copier On? Energy Star Copier Performance* (LBNL-41332; San Francisco, CA: Lawrence Berkeley National Laboratory, February 1998).
8 EPA Climate Protection Partnerships Division, *2000 Annual Report.*
9 *Ibid.*

EPA is not resting on its laurels. In 2003 and the years ahead, Energy Star will be adding new products that can earn the label and will be building consumer awareness of the Energy Star label as the national and global symbol of energy efficiency. Energy Star is expanding its labeling and recognition programs in the building and industrial sectors, and is expanding and strengthening its voluntary partnerships.

Energy Star time-line

1989 ● ICOLP founded by EPA and 14 companies in the electronics and aerospace industries. This marks the beginning of EPA voluntary partnerships.

1991 ● EPA introduces "Green Lights," a partnership program to promote efficient lighting in commercial and industrial buildings (to be integrated in Energy Star by the end of the decade).

1992 ● First Energy Star-labeled products including personal computers and monitors.

1993 ● Energy Star extended to printers.
● AgSTAR voluntary program to reduce methane emissions.
● EPA/DOE "Climate Wise" corporate leadership initiative.
● Executive Order directs US agencies to purchase only Energy Star desktop computers, monitors, and printers and asks private organizations to do the same.

1994 ● Energy Star extended to fax machines.
● EPA Coalbed Methane Outreach Program.

1995 ● Energy Star extended to copiers, transformers, residential heating and cooling products (including air-source heat pumps, central air conditioning and furnaces, gas-fired heat pumps, and programmable thermostats).
● Green Lights merged in Energy Star Buildings.
● Energy Star Residential.
● Energy Star Transformer Program.
● Natural Gas Star.
● EPA Voluntary Aluminum Industrial Partnership (VAIP).

1996 ● DOE joins EPA in Energy Star partnership.
● Energy Star extended to exit signs, insulation, and appliances (including dishwashers, refrigerators, and room air conditioners).

1997 ● Energy Star extended to clothes washers, light fixtures, multifunction office equipment, and scanners.

1998 ● Energy Star extended to TVs and VCRs.

1999 ● Energy Star extended to consumer audio and digital versatile disk (DVD), roof products, and compact fluorescent lights.
● Energy Star Commercial Buildings.

2000
- Energy Star extended to water coolers and traffic signals.
- US Army and Navy home procurement requires Energy Star specifications
- Energy Star Schools.

2001
- Energy Star extended to set-top TV boxes, residential dehumidifiers, ceiling fans, ventilation fans, telephone products, and commercial washing machines.
- Energy Star Supermarkets and Hospitals.
- Energy Star Hotels.

5

Japan's F-Center for developing greenhouse gas alternatives

Sustainable living through better chemistry*

The first research center devoted to "sustainable chemistry" is designing chemicals that are safe for the ozone layer, climate, and ecosystems, while working in collaboration with chemical industries and the Japanese Ministry of Economy, Trade, and Industry.

When Nelson Mandela took over the reins of government in South Africa, he prophesized: "If globalization is to create real peace and stability across the world, it must be a process benefiting all."

Such was the motivation and the inspiration in April 2001, when the new Greenhouse Gas Alternatives Center began its operations. Officially named the Research Center for Developing Fluorinated Greenhouse Gas Alternatives, or the "F-Center," it is located in the National Institute of Advanced Industrial Science and Technology (AIST) in Tsukuba, Japan. This organization is playing an important role as a "center of excellence" for sustainable chemistry.

The F-Center[1] is the first research institute devoted to sustainable chemistry. Sustainable chemicals are those that are safe to plants and animals, have short atmospheric lifetimes, and have an environmentally benign atmospheric fate—in other words, no permanent or non-reversible environmental effects. The F-Center has helped to develop new hydrofluoroethers (HFEs) that replace CFCs in many

* The authors are grateful for interviews and supplementary assistance by AIST scientists and managers Shuzo Kutsuna, Akira Sekiya, Masaaki Sugie, Masanori Tamura, Katsumi Tanaka, Tadafumi Uchimaru, Sachiko Uehara, and Masaaki Yamabe; by Government of Japan experts Kouichiro Kakee, Naoki Kojima, Haruhiko Kono, Masahiro Miyazaki, Tetsuo Nishide, Toshiki Sakurai, Yoshihiko Sumi, and Takashi Ueda; and by Research Institute of Innovative Technology for the Earth (RITE) managers Bunji Amemiya and Susumu Misaki.

1 Hereafter in this book, the Research Center for Developing Fluorinated Greenhouse Gas Alternatives is referred to as the "F-Center."

Figure 5.1 The Research Center at the National Institute of Advanced Industrial Science and Technology (AIST) in Tsukuba

Members of the Research Center together with Stephen Andersen, Helen Tope, and Yuichi Fujimoto

Front (left to right): Kanako Sato and Yuichi Fujimoto; *middle* (left to right): Masaaki Yamabe, Katsumi Tanaka, Stephen Andersen, Helen Tope, Akira Sekiya, Masanori Tamura, and Shuzo Kutsuna; *back* (left to right): Heng-dao Quan, Tadafumi Uchimaru, Liang Chen, Kazuaki Tokuhashi, Shigeo Kondo, Takashi Abe, and Yasuhisa Matsukawa

applications while satisfying these demanding criteria. The center is also developing energy-saving processes to produce the most sustainable chemicals.

AIST's philosophy is quite simple as well as very enlightened: follow the triple bottom line—people, planet, and profits. Now, and for all future generations.

The F-Center philosophy has led to a new approach to the development of alternatives to CFCs. This is important because scientists have continually discovered serious unanticipated environmental consequences of manufactured chemicals. The F-Center chooses environmental sustainability as its first design criterion—and this is a complete paradigm shift. To understand the uniqueness of the center's approach to the development of such chemicals, it is important to appreciate the history of chemical development. Figure 5.2 shows the F-Center's strategy for developing sustainable chemicals.

Chemistry is older and more important than you think

Throughout human history, chemicals have been essential to prosperity. Chemicals are used to preserve food, to cure our illnesses, to control hazardous and harmful pests, to create durable and recyclable materials, and to protect workers against disease and hazards. Chemicals—in the form of plastics—are also essential to produce the solar cells and fuel cells that will diminish dependence on fossil fuel in the future.

Today the value of world chemical production exceeds $1.7 trillion annually, of which almost 30% is traded internationally. Growth in the demand for chemicals is projected to continue at 2.4% per year among the developed countries, and at 5.9% in developing countries. Chemicals are produced in nearly every country, employing over 10 million people in their manufacture.[2]

But chemicals can be deadly and environmentally destructive when misused or mismanaged.

At first, humans could only see the benefits of chemistry and overlooked the unintended consequences of using chemicals. Awareness of the dangers of chemicals does reach far back into history, though. The most acutely toxic, naturally occurring chemicals were known even in the earliest periods of Roman, Greek, and Chinese civilization; and by the 15th century, these natural toxic chemicals were commonly used as pesticides and weapons.[3]

---

2 United Nations Environment Program (UNEP), *Industry as a Partner for Sustainable Development Ten Years after Rio: The UNEP Assessment* (Paris: UNEP, 2002).

3 In the earliest periods of Roman, Greek, and Chinese civilization the most obvious human and animal poisons were categorized for use as pesticides and for warfare and assassination, spawning the profession of royal "tasters." And 2,400 years ago Socrates became a celebrated poisoning victim when he was executed by poison hemlock. Early chemical industrialization often had serious consequences. For example, poisoning from the lead used for food and wine vessels and plumbing was a contributing factor in

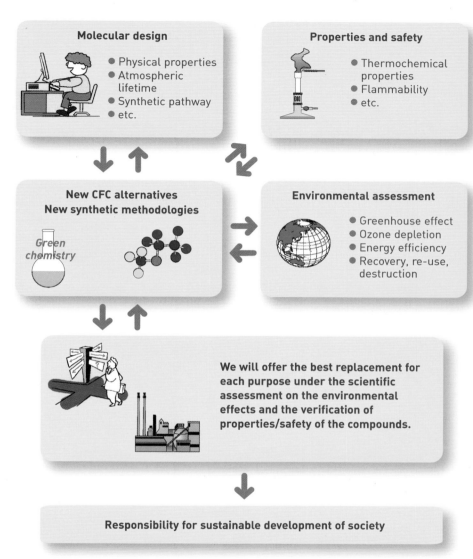

Figure 5.2 F-Center strategy for developing sustainable chemicals

the decline of the Roman Empire. During the 15th century the city council of Venice, Italy, kept detailed records of assassination victims, contractor, cost, poison type, and dose. Lucrezia Borgia (1481–1519) supposedly killed family members, lovers, political opponents, and churchmen; and Catherine de Medici (1519–89, wife of Henry II of France) was an early "experimental toxicologist" who poisoned poor and sick street people and recorded the symptoms and dose-response. See http://toxicology.usu.edu/ 660/html/history.htm.

When the first manufactured chemicals were developed in the late 1800s and early 1900s, there was little understanding of their potential carcinogenic, reproductive, or ecosystem-harming effects. The new chemicals brought high profits to inventors and manufacturers; and their direct benefits to society were extraordinary. For example, national governments and health authorities quickly embraced dichlorodiphenyltrichloroethane (DDT) and similar pesticides to combat diseases such as malaria that were carried by insects. DDT was considered a miracle pesticide, worthy of the Nobel Prize. It was not until the 1950s and 1960s that people began to realize that some chemicals cause cancer; but in those times scientists were still not always able to determine the exact chemical cause of the cancers.[4]

As late as 1960, few regulations restricted or prevented the manufacture and sale of chemicals hazardous to humans, and there was even less concern for environmental and indirect health effects. However, the outlook changed dramatically after the publication in 1962 of Rachael Carson's *Silent Spring*,[5] which documented the widespread ecological and health damage of pesticides.

In the four decades since *Silent Spring*, chemical companies and the public worldwide have become more concerned and educated. The first environmental ministries have been created and given unprecedented authority, and the first international environmental protection treaties have been signed. This rapid progress was stimulated by persuasive scientific proof of the ecological and health effects of radiation, asbestos, air and water pollution, and toxic waste disposal.

The dawn of global environmental awareness

The first environmental agencies were created around 1970. By 1975, most developed-country governments had reorganized themselves to put a priority on environmental protection. Then, in just a decade, the world experienced five catastrophic ecological and engineering disasters that lead to prompt investor, public, and regulatory action: Three Mile Island (1979), Bhopal (1984), the discovery of the Antarctic ozone hole (1985), Chernobyl (1986), and the *Exxon Valdez* (1990).

At the same time, dramatic progress occurred in health and environmental science and in the technology needed to monitor the effects of chemicals. Rapid evolution in laboratory equipment used to detect chemical contamination made it possible to link illnesses to exposure to certain chemicals. This improvement in technology had begun back in 1956, when Dr. James E. Lovelock invented the electron-capture detector, a device for gas chromatography that could detect tiny

---

4   Beginning in the 1700s, scientists noted occupational health effects of many chemicals but did not determine the specific cause. For example, the debilitating neurological damage called "mad-hatter's disease" was caused by mercury used in the felting of hats, and "plumber's syndrome" was caused by lead used in pipes. In the early 1800s, Spaniard V. Orfila (1787–1853) became the "father of modern toxicology," compiling the first chemical and biological information on most known poisons and proposing the necessity of chemical analysis to prove cause and effect.

5   Rachael Carson, *Silent Spring* (Boston, MA: Houghton Mifflin, 1962).

amounts of chemical compounds in the atmosphere and on Earth. This device, which is one million times more sensitive than conventional thermal conductivity detectors, made it possible to detect halogenated compounds at levels of only one part per trillion, thus revolutionizing our understanding of the atmosphere and pollutants.[6]

By the 1970s, scientists were using the electronic-capture detector and other technologies to prove that many chemicals caused cancer; and health and environmental regulators began removing the most dangerous chemicals from the market. At the same time, scientists documented that some chemicals caused more subtle health effects, including birth defects and mutation.[7]

Building on this remarkable progress, scientists continue today to sound the alarm as unanticipated and unnoticed effects of chemicals designed to benefit humans turn out instead to threaten prosperity and even survival. CFCs are a case in point.

When CFCs were invented in the 1930s, they were considered a miracle chemical.[8] It was not until 1974 that Mario Molina and Sherwood Rowland hypothesized that CFC emissions could deplete the stratospheric ozone layer.[9] And it took until the years between 1985 and 1987 for countries to reach a global consensus on how to address this problem. The effects of CFC emissions are indirect, because, while it is generally safe to be exposed to CFCs directly, they destroy the ozone layer that shields the Earth from the sun's ultraviolet radiation (Figure 5.3). Increased exposure to ultraviolet radiation, in turn, increases skin cancer and cataracts, suppresses the human immune system, and reduces productivity of agricultural and natural ecosystems.

Similarly, it was not until the 1990s that ecologists and health scientists warned that some "persistent" organic chemicals remained in ecosystems for hundreds, or even thousands, of years—and that at least a dozen of these chemicals cause potentially serious biological effects. A new concern emerged as well: that chemicals could harm human health in unexpected ways by stimulating an immune response, by altering cell function, or by interference with hormones. Another problem caused by chemicals with extremely long lifetimes was global climate change, accelerated by fossil fuel combustion and emissions of manufactured gases.

The development of CFCs and persistent organic chemicals is typical of the historic course of chemical development—which has typically been mass produc-

6 James Lovelock, "Award Lecture of the Blue Planet Prize 1997," Asahi Glass Foundation, Japan.

7 The famous book *Our Stolen Future* (T. Colborn, D. Dumanoski and J.P. Myers, *Our Stolen Future* [New York: Dutton, 1996]) explains the particulars of how the numerous persistent synthetic chemicals disposed widely around the world have been concentrated via the food chain and introduced into our bodies, disrupting the endocrine systems of both wildlife and humans. See also Environmental Working Group, http://ewg.org.

8 For the history of ozone-layer protection, including the importance of industry leadership, see Stephen O. Andersen and K. Madhava Sarma, *Protecting the Ozone Layer: The United Nations History* (London: Earthscan Publications, 2002).

9 M. Molina and F.S. Rowland, "Stratospheric Sink for Chlorofluoromethanes: Chlorine Atom-Catalyzed Destruction of Ozone," *Nature* 249.5460 (1974): 810-12.

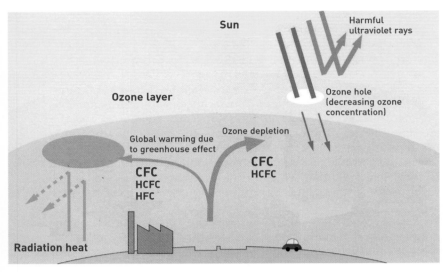

Figure 5.3 Impacts of manufactured chemicals on ozone and climate

CFCs, HCFCs, and HFCs have convenient numbered acronyms. When reading a code from right to left, the first digit defines the number of fluorine atoms; the second is the number of hydrogen atoms *plus one*; and the third is the number of carbon atoms *minus one* (see Table 5.1 on page 96 for examples).

tion first, scientific discovery of adverse effect second, and then action to abandon the chemical.

This self-defeating attitude ultimately prompted the F-Center to take a new, preventative approach.

A revolutionary new approach to chemistry

The F-Center's new, sustainable approach to developing alternatives to CFCs considers the life-cycle effects of chemicals in the early stages of development (see Figure 5.4). It does not wait for production of chemicals to begin before researching harmful effects.

This thinking is really quite revolutionary. The F-Center's first consideration is the possible environmental impact of the alternative CFCs, such as their contributions to climate change. The organization's second consideration is the potential usefulness and energy-saving value of the alternative chemical. Currently, the F-Center is investigating new fluorine chemistry and using computational chemistry to develop environmentally benign molecular structures and preferred synthetic routes. A director of the F-Center, Masaaki Yamabe, explains:

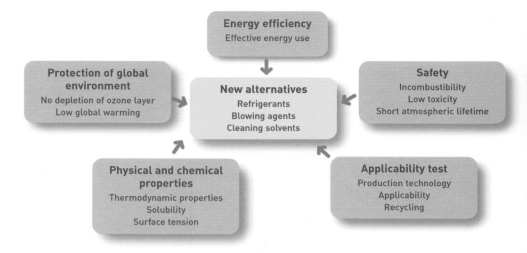

Energy efficiency
Effective energy use

Protection of global environment
No depletion of ozone layer
Low global warming

New alternatives
Refrigerants
Blowing agents
Cleaning solvents

Safety
Incombustibility
Low toxicity
Short atmospheric lifetime

Physical and chemical properties
Thermodynamic properties
Solubility
Surface tension

Applicability test
Production technology
Applicability
Recycling

Figure 5.4 The challenge of integrating environmental protection

By the time that the Montreal Protocol was agreed upon, the stratospheric ozone layer was already so jeopardized that an emergency phaseout of the most potent ozone-depleting substances was necessary. Chemical companies found substitutes that could be commercialized rapidly and adopted by customers, but with the disadvantage that they are mostly hydrochlorofluorocarbons (HCFCs) or hydrofluorocarbons (HFCs). These chemicals have allowed us to save the ozone layer, but some of them are causing climate change. Now society needs either to eliminate those global-warming compound emissions in applications where they provide no environmental benefit, or to replace them with superior technology.

Sustainable chemistry, first implemented by the F-Center, is the synthesis of earlier approaches including "green chemistry," "responsible use," "pollution prevention," "100% product," "life-cycle climate performance," and "reversibility." Achieving it is a very tough challenge, but the engineers at F-Center know that nothing really useful for a sustainable society will result without accepting a challenge.

Green chemistry encourages the development and use of natural and manufactured chemicals that are inherently safe and environmentally acceptable. **Responsible use** is a philosophy of choosing the most environmentally acceptable technology to satisfy product, safety, and environmental performance, and to contain, recover,

Masaaki Yamabe, a director of the F-Center, with ozone champion poster

and re-use chemicals to minimize emissions. **Pollution prevention** is the realization that it is often both economically and environmentally superior to prevent pollution rather than to suffer its consequences and pay later for the cost of proper disposal. Similarly, **100% product** is a manufacturing goal of turning all inputs into products, thereby avoiding the expense and environmental effects of disposal of inputs that are otherwise turned to waste. **Life-cycle climate performance** is the accounting of all direct and indirect emissions of greenhouse gases over a complete life-cycle.[10] **Reversibility** is the ability of the atmosphere and ecosystem to recover quickly from the emission of a chemical.

Sustainable chemistry, as applied by the F-Center, ties all of these approaches together. The F-Center may be the only organization in the world that puts together the full range of approaches and also a comprehensive range of institutions for the purpose of testing and developing better chemicals.

A concern for future generations

Ozone-depleting substances and greenhouse gases were developed to serve important human needs such as refrigeration of food, aerosol delivery of medicine, and human safety. They are also used for convenience, cosmetic, and frivolous applications such as noisemakers, party foam dispensers, and self-chilling beverages. Figure 5.5 shows some of these applications.

Says Yamabe:

> It is natural that companies that worked so hard to phase out CFCs would be defensive about proposals to limit the emissions of HCFCs and HFCs. But Japanese companies realize a strong obligation to future generations. This is rooted deep into our culture. We place a high value on the family and on our society. We also have much confidence in our engineering and chemistry. The challenge for chemists is to listen carefully to atmospheric scientists and ecologists and to satisfy environmental performance criteria.
>
> One reason Japan is leading on sustainable chemistry is that we appreciate the advantage of cooperation. My introduction to ozone-layer protection was in 1986. Dr. Nobuo Ishikawa and I were selected for the US EPA Experts Panel on CFC Alternatives, an international committee investigating the economic feasibility of chemical alternatives. We were honored to serve with experts from Germany, Italy, Japan, the United Kingdom, and the United States.[11]

Dr. Ishikawa was a Professor of Chemistry at the Tokyo Institute of Technology and Chairman of the Industry Advisory Committee (Kagakuhin Shingikai) in the Ministry of International Trade and Industry (MITI). He was also a trusted advisor

10   Including the emissions of synthesis of chemicals, direct and chemical by-products, recovery, and destruction.

11   The panel of experts was formed in February 1987, and issued its report on May 24, 1987, concluding that the absence of a market—rather than technical or environmental issues—was the principal barrier to commercialization of CFC substitutes.

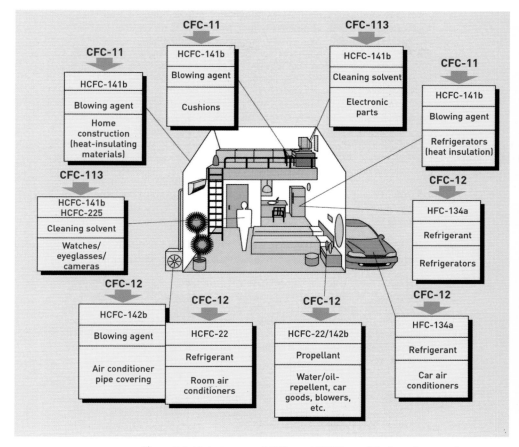

CFC-11
⬇
HCFC-141b
Blowing agent
Cushions

CFC-113
⬇
HCFC-141b
Cleaning solvent
Electronic parts

CFC-11
⬇
HCFC-141b
Blowing agent
Home construction (heat-insulating materials)

CFC-11
⬇
HCFC-141b
Blowing agent
Refrigerators (heat insulation)

CFC-113
⬇
HCFC-141b HCFC-225
Cleaning solvent
Watches/ eyeglasses/ cameras

CFC-12
⬇
HFC-134a
Refrigerant
Refrigerators

CFC-12
⬇
HCFC-142b
Blowing agent
Air conditioner pipe covering

CFC-12
⬇
HCFC-22
Refrigerant
Room air conditioners

CFC-12
⬇
HCFC-22/142b
Propellant
Water/oil-repellent, car goods, blowers, etc.

CFC-12
⬇
HFC-134a
Refrigerant
Car air conditioners

Figure 5.5 Historic uses of CFCs and HCFCs, and their replacements

to the People's Republic of China and the Union of Soviet Socialist Republics. At that time, Yamabe was director of fluorine chemistry at Asahi Glass—then a prominent Japanese manufacturer of CFCs and HCFCs.

Yamabe reminisces about his time on the committee.

> We were all very proud to be on this committee. Of course we knew of the other famous chemists, but we hadn't had the opportunity to actually work together. Imagine the intellectual power of the dozen best fluorine chemists in the world! We could do anything! We started by identifying all of the chemicals that first replaced CFCs, including HCFC-225, the invention of chemical engineers at Asahi Glass.

He continues:

> The successful development of HCFC-225 eventually brought me particular pleasure as the project leader. But I heard an interesting story from a

government official in the USA. When we published the EPA report listing HCFC-225 as an environmentally superior substitute for CFC-113, DuPont—Asahi Glass's arch-rival—went out of its way to belittle us, saying that HCFC-225 might be a great solvent, but that it would require such a sophisticated catalyst that it could never be manufactured. At that time, they probably feared that confidence in CFC substitutes would only encourage their regulation.

And I am sure they also suffered from the "not-invented-here syndrome," since they were first to discover CFCs. We quickly developed a trail-blazing catalyst and completed our pilot manufacturing facility, just as DuPont turned up the volume on claims that CFC-113 could never be replaced. In the end, most CFC-113 customers chose no-clean soldering or aqueous solutions [see Chapter 10 on Visteon], so not much HCFC-225 was ever sold. But the option of HCFC-225 helped environmental leadership in companies like AT&T, Nortel, and Seiko Epson to have the confidence to announce the CFC phase-out. I am very proud that Asahi Glass made this strategic contribution.

By 1987, DuPont was reconsidering its opposition to ozone-layer protection, and it soon became a leader in the Montreal Protocol phase-out. Yamabe says:

Today, DuPont-Mitsui Fluorochemical Company is one of the supporting members of the national project that is striving to commercialize next-generation chemicals in collaboration with chemical industries and the F-Center, under the sponsorship of the new Ministry of Economy, Trade, and Industry (METI).[12] Companies that did not have the experience of cooperation for ozone-layer protection simply could not imagine that what we did was possible.

How the Montreal Protocol really works

The Montreal Protocol bans the continued production of CFCs, and later bans the production of transitional substances such as HCFCs. Developed countries are allowed to produce HCFCs for new equipment until 2020, and for service until 2030. However, some developed countries plan to stop HCFC-22 production for new air-conditioning equipment by 2010, and to stop its production for service by 2020. Adequate quantities of HCFC refrigerants can be supplied for the economic life of equipment by recycling and stockpiling.

HCFC-22 is a high-pressure gas with a Montreal Protocol ozone-depletion potential (ODP) of 0.055 and a global-warming potential (GWP) of 1,700. HCFC-123 and HCFC-225 are low-pressure gases with a Montreal Protocol ODP of 0.02 and 0.02–0.03, and a direct GWP of 120 and 170–530, respectively.

---

12   Japan reorganized MITI and renamed it METI in January 2001. The government champions of the MITI and METI include: Naoki Kojima, Yoshihiko Sumi, Tetsuo Nishide, Masahiro Miyazaki, Takashi Ueda, Toshiki Sakurai, Haruhiko Kono, and Kouichiro Kakee.

Some HCFCs, like HCFC-123 and HCFC-225, have both very low ODPs and GWPs (see Table 5.1); but other HCFCs are not environmentally sustainable due to their higher ODPs and GWPs. HFCs are greenhouse gases that are controlled by the Kyoto Protocol, but HFCs are ozone-safe and generally have a GWP much lower than the ozone-depleting substances (ODSs) they replace.[13]

HFCs, with a shorter lifetime, have less impact on climate change than the CFCs they replace, but are still included in the basket of greenhouse gases whose emissions need to be reduced. Some countries, including the USA and Japan, will use voluntary and negotiated measures to encourage energy efficiency and to minimize HFC emissions. Other countries plan to restrict HFC use and emissions more directly. HFC blends, including HFC-407C and HFC-410A, are often selected to optimize the energy performance of specific air-conditioning systems and particular sizes of units. HFC blends have zero ODP and a GWP that depends on the exact composition.[14]

In the majority of cases, the replacement of ODSs has been accomplished with no compromise in any measure of environmental risk and safety performance. The new technology is generally considered to be environmentally acceptable. Approximately 80% of CFCs and halon that would be used in developed countries today, had the Montreal Protocol not been developed, have been successfully phased out without the use of other fluorocarbons such as HCFCs or HFCs. Developing countries have replaced 76% of ODSs with non-fluorocarbons. This was done through a combination of "not-in-kind"[15] chemical substitutes, product alternatives, manufacturing-process changes, conservation, and simple avoidance.

Non-CFC fluorocarbons (HFCs and HCFCs) have often been the substitutes in both developed and developing countries, with HFCs substituting for 8% and 7% in developed and developing countries respectively and HCFCs substituting for 12% in developed countries and 17% in developing countries. Approximately 76–80% of ozone-depleting compounds used by developing and developed countries respectively have been successfully replaced without the use of other fluorocarbons (see Table 5.2). By 2000, the global production of CFCs was less than 4% of its peak production level of 1988.

HCFCs and HFCs, though environmentally superior to the CFC chemicals they replace, are both scheduled for further environmental regulation. HCFCs are necessary as "transitional substances" in applications where completely ozone-safe alternatives and substances are not yet available or are not technically or

13   ODPs and GWPs are available from several sources: Montreal Protocol, *Scientific Assessment of Ozone Depletion 1998*; IPCC (Intergovernmental Panel on Climate Change), *Third Assessment Report: Climate Change 2001*; and the *Handbook for the International Treaties for the Protection of the Ozone Layer* (2000 edition). Note that the Montreal Protocol ODPs were officially established in 1992 and may be reviewed and revised periodically. The 2002 *Scientific Assessment* presents the World Meteorological Organization 1999 Model results as shown in Table 5.1 overleaf.

14   For a comprehensive comparison of the direct emissions and the indirect emissions from energy consumption, see UNEP Technology and Economics Assessment Panel (TEAP), *The Implications to the Montreal Protocol of the Inclusion of HFCs and PFCs in the Kyoto Protocol* (Nairobi, Kenya: UNEP, 1999).

15   "Not-in-kind" substitutes are chemical compounds without any fluorine atom in a molecule, or are non-chemical alternatives.

	Name of substance	Chemical formula	Boiling point [°C]	Ozone-depletion potential (ODP)	Global-warming potential (GWP [100])	Atmospheric lifetime (year)	Refrigerant	Blowing agent	Cleaning solvent
CFC	CFC-11	CCl_3F	23.7	1.0	4,600	45	✓	✓	
	CFC-12	CCl_2F_2	−29.8	0.82	10,600	100	✓	✓	
	CFC-113	CCl_2FCClF_2	47.6	0.90	6,000	85			✓
	CFC-114	$CClF_2CClF_2$	3.6	0.85	9,800	300	✓		
	CFC-115	CF_3CF_2Cl	−38.9	0.40	7,200	1,700	✓		
HCFC	HCFC-22	$CHClF_2$	−40.8	0.034	1,700	11.9	✓	✓	
	HCFC-123	CF_3CHCl_2	27.7	0.012	120	1.4	✓	✓	✓
	HCFC-141b	CH_3CCl_2F	32.1	0.086	700	9.3	✓	✓	✓
	HCFC-142b	CH_3CClF_2	−9.8	0.043	2,400	19.0		✓	
	HCFC-225ca	$CF_3CF_2CHCl_2$	50.1	0.025	180	2.1			✓
	HCFC-225cb	$CClF_2CF_2CHCl$	53.4	0.033	620	6.2			✓
HFC	HFC-23	CHF_3	−82.2	0	12,000	260	✓		
	HFC-32	CH_2F_2	−51.7	0	550	5.0	✓		
	HFC-125	CHF_2CF_3	−48.5	0	3,400	29	✓		
	HFC-134a	CH_2FCF_3	−26.1	0	1,300	13.8	✓	✓	
	HFC-143a	CH_3CF_3	−47.2	0	4,300	52	✓		
	HFC-152a	CHF_2CH_3	−24.1	0	120	1.4	✓		
HFE	HFE-245mc	$CF_3CF_2OCH_3$	5.5	0	530	4.5	✓		
	HFE-143m	CF_3OCH_3	−24	0	680	4.9	✓		
	HFE-347mcc	$CF_3CF_2CF_2OCH_3$	34.2	0	490	4.6	✓	✓	
	HFE-347mmy	$(CF_3)_2CFOCH_3$	29.4	0	350	3.5	✓		
	HFE-347pc-f	$CHF_2CF_2OCH_2CF_3$	63.8	0	870	6.0			✓
	HFE-356mec	$CF_3CHFCF_2OCH_3$	54.3	0	N/A	1.5			✓
	HFE-55-10mec-fc	$CF_3CHFCF_2O\text{-}CH_2CF_2CHF_2$	105.9	0	N/A	5.0			✓
	HFE-254pc	$CHF_2CF_2OCH_3$	37.2	0	330	2.4	✓		

Table 5.1 Ozone-depletion potential (ODP) and global-warming potential (GWP)

Source: IPCC Third Assessment Report: Climate Change 2001

	Portion not fluorocarbons	Portion fluorocarbons	
		HCFCs	HFCs
Developed countries	80%	12%	8%
Developing countries	76%	17%	7%

Table 5.2 The replacement of CFCs

economically feasible. HCFCs are chemicals with relatively low ODP and GWP, but the Montreal Protocol does call for their use to be halted eventually.[16] HFCs are necessary for applications where toxic and flammable alternatives to CFCs cannot be safely used. HFCs are ozone-safe chemicals with lower GWPs than the CFCs they replace.[17]

How government became involved

In 1989, the Senate of Japan asked MITI to strongly promote development of chemical alternatives to ozone-depleting substances (ODSs). In the meantime, Dr. Akira Sekiya, who was head of a fluorine chemistry laboratory in the National Institute of Materials and Chemical Research (NIMC), realized the importance of this proposal, and started basic research in his group even before funding was awarded.

In 1990, MITI moved quickly to finance this research, and also decided to start the first national five-year project for development of "advanced refrigerant for compression heat pumps." The government grant was awarded to the RITE through the New Energy and Industrial Technology Development Organization (NEDO). The RITE organized a research consortium with ten companies, and contracted with NIMC (Sekiya Group) for a joint venture. This project focused research on refrigerants. In 1992, the team increased its environmental capability in collabora-

---

16 For developed countries, the Montreal Protocol allows production of HCFCs for use in new products until 2020 and production for the servicing of those products until 2030. For developing countries, the Montreal Production allows production until 2040.

17 The Montreal Protocol created the treaty precedent of "baskets" of chemically similar compounds that are to be reduced in any combination to achieve the desired overall reduction in ozone-depleting potential. The Protocol defines a base year for each category and multiplies the ODP by the total quantity of chemicals produced in that year to establish "total ODP-weighted tonnes." In subsequent control periods, the allowed ODP-weighted tonnes for each category (specified as a percent reduction from the base year) can be composed of any combination of ODSs. Similarly, the Kyoto Protocol created a basket of six greenhouse gases (CO_2, methane, nitrous oxide, HFCs, PFCs, and SF_6) that can be reduced in any combination to satisfy the agreed reduction in weighted greenhouse gas emissions.

Dr. Akira Sekiya, Deputy Director
of the F-Center

tion with the National Institute for Resources and Environment to evaluate the atmospheric lifetime and global-warming potential of the developed chemicals.

In 1992, Tetsuo Nishide was designated head of the Ozone Layer Protection Policy Office and took over management of the ozone-protection project at MITI. This was two years before the completion of the first project, but he immediately decided to expand the scope of the program and started preparation of the second phase project. This was the development of "new refrigerants and other substances for effective use of energy," and Nishide invited the research group at the National Industrial Research Institute of Nagoya to join in. This second phase of the research came to completion in March 2002, and it centers on several HFEs.

Nishide is a champion of environmental protection and an inspired research manager who has a willingness to take risks. "Skeptics of government research predicted that commercially attractive chemicals would be developed by private companies or that the environmental criteria were too stringent and no research group would identify acceptable chemicals," said Nishide. Despite this view, MITI pursued the projects that transformed the thinking of chemical companies.

The national project is a consortium of fiercely competing multinational companies that have Japanese operations (see Table 5.3). These companies collaborated

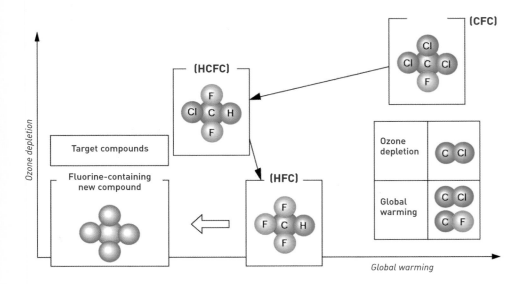

Figure 5.6 Molecular structure of CFCs, HCFCs, HFCs, and new fluorine-containing compounds

Company	Phase 1: fiscal years 1990–1994	Phase 2: fiscal years 1994–2002
Asahi Chemical Industry	✓	✓
Asahi Glass	✓	✓
Central Glass	✓	✓
Daikin Industries	✓	✓
DuPont–Mitsui Fluorochemicals	✓	✓
Kanto Denka Kogyo	✓	✓
Showa Denko K.K.	✓	✓
Toagosei Company	✓	✓
Tohkem Products Corporation	✓	
Tosoh Corporation	✓	

Table 5.3 Company participation in the F-Center ODS replacement project

with industries and public research organizations (such as NIMC and F-Center) to find environmentally superior, next-generation replacements for such applications as refrigerants, insulating foam-blowing agents, and solvents.

During the first and second phases of research, from 1990 to 2002, participating companies served on a coordinating committee at RITE and dispatched company research scientists to work together with staff scientists of NIMC in Tsukuba during three- to five-year rotations. Academic scientists were also contracted from Keio University, Kobe University, Nihon University, Ibaraki University, Sophia University, the Japan Industrial Conference on Cleaning, and the US Atmospheric Environmental Research.

Nishide said:

> When we apply these strong environmental chemical design criteria, we naturally view the HFCs as the commercial benchmark, since they are the most common fluorocarbon today. We want our new chemicals to be better in every way. But, for refrigerants, we also find ourselves competing with some of the oldest chemicals like carbon dioxide and propane, and even with more toxic chemicals like ammonia.

How and why older chemicals are making a comeback

Carbon dioxide, propane, butane, and ammonia—originally abandoned or in limited use due to safety reasons—are coming back into use as refrigerants because engineers have invented ways to use them more safely (see Figure 5.7 and Chapter 3 on DaimlerChrysler CO_2 vehicle air conditioning). Carbon dioxide is very hard to contain in refrigeration equipment because it has operating pressures more than six times higher than other refrigerants. Propane and butane are highly flammable. Ammonia is flammable, toxic, and also pungent (causing burning eyes and throat when breathed even at non-toxic concentrations). It is no wonder these were not considered good refrigerants until engineering marvels of the last decade created reliable systems to contain them. But old reputations die hard, and it has taken some innovative marketing to get the public to accept these refrigerants.

Greenpeace has been very successful in promoting propane, butane, and even pentane for the small-size refrigerators popular in Europe; and that technology is spreading now to China, India, Japan, and elsewhere. Small refrigerators use a small refrigerant charge that can be kept away from ignition sources. Propane

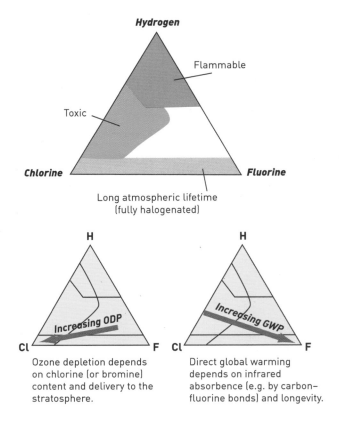

Figure 5.7 Toxicity, flammability, ozone, and climate trade-offs in chemical design

Source: James M. Calm and David A. Didion, "Trade-Offs in Refrigerant Selections: Past, Present, and Future," paper presented at Refrigerants for the 21st Century, American Society of Heating, Refrigeration and Air-Conditioning Engineers (ASHRAE)/National Institute of Standards and Technology (NIST) Refrigerants Conference, Gaithersburg, MD, October 6–19, 1997.

refrigerants have very good energy efficiency, and Greenpeace played a critical role in persuading consumers to accept the new technology despite its higher fire risk. Greenpeace has also played a positive role in working with safety authorities to make sure that manufacturers take all the steps necessary for safety.

The engineering challenge is to ensure adequate safety in larger refrigeration and air-conditioning equipment that requires a greater quantity of refrigerant. So far, safety authorities have often been reluctant to allow big systems to use flammable and toxic refrigerants, and companies are reluctant to face the risk of liability.

Engineers "mitigate" the risks of hazardous chemicals through standard strategies: reduce the quantity of chemical to the smallest workable amount, contain the chemical carefully to avoid exposure to humans or fire, remove open sources of ignition by isolating and sealing electric switches, and install detectors that initiate venting or sound alarms.

> **STEPS TOWARD ENVIRONMENTAL IMPROVEMENT**
>
> *Step 1* Evaluate methods of chemical discovery and molecular design.
>
> *Step 2* Screen all possible candidate substances and select those with a clear environmental advantage.
>
> *Step 3* Synthesize alternative substances, measure and evaluate safety, toxicity, atmospheric lifetime, ozone-depletion potential, atmospheric fate, and necessary properties for use as refrigerants, blowing agents, and cleaning solvents.
>
> *Step 4* Evaluate application tests, manufacturing feasibility, environmental acceptability, and cost estimates, and perform overall evaluation of market potential and environmental merit.

This safety mitigation can sometimes, but not always, be accomplished without any adverse effect on energy performance. For example, detection systems, ventilation systems, and fireproof electrical systems do not affect energy efficiency. However, some safety systems, like a secondary cooling loop, do reduce energy efficiency. One example is a large hydrocarbon refrigeration system that is placed in an isolated outside location, with cold water pumped into buildings for cooling. Such systems are increasingly popular for the refrigeration of food in European supermarkets. The pumping, and the energy lost as the water travels through long pipes, consumes energy and so increases emissions of greenhouse gases from power plants, at least partially offsetting the advantage of using the low-GWP refrigerant.

All of the project scientists at F-Center agree that sustainable chemistry provides a way to develop the chemicals that humans need while avoiding the health costs and economic losses that came from the older methods of developing chemicals first and then learning about their effects later. The F-Center's forward-thinking approach is enabling it to invent the menu of chemicals that will allow future generations to enjoy a prosperous, healthy, and safe world.

SUSTAINABLE CHEMISTRY ENVIRONMENTAL PARTNERSHIPS

The International Council of Chemical Associations is working towards "Responsible Care®,"* along with the UNEP, the World Business Council for Sustainable Development, the International Chamber of Commerce, and the Intergovernmental Forum on Chemical Safety. Responsible Care calls for no accidents, injuries, or harm to the environment.

The Alliance for Responsible Atmospheric Policy, UNEP, and the US EPA as well as the Japanese METI and industry associations such as Japan Industrial Conference for Ozone Layer Protection (JICOP) are organizing a complementary HFC Responsible Use Partnership. The partners exercise chemical stewardship of HFCs only where necessary for the successful phase-out of ozone-depleting substances or when they provide health and safety, environmental, technical, economic, or unique society benefits. In addition, partners make every technical and economic effort to minimize emissions.

At the F-Center, these approaches are translated into molecular design criteria for chemicals as follows:

Protection of the global environment

- Very short atmospheric lifetime (preferably less than ten years)
- Zero depletion of ozone layer (based on Montreal Protocol-ODP or World Meteorological Organization-ODP)
- Low global warming impact estimated by some modified LCCP, which is even more comprehensive than the original calculation of total equivalent warming impact (TEWI)†
- Benign and reversible atmospheric fate (no persistent hazardous compounds resulting from decomposition)

Energy efficiency

- Energy efficiency greater than achievable with existing chemicals

Safety

- Low or mitigated flammability
- Low or mitigated toxicity

Physical and chemical properties

- Refrigerants
- Thermodynamics matched to refrigeration cycle (different refrigerants perform more efficiently in certain refrigeration or A/C systems)
- Chemical stability, materials and lubricant compatibility
- Insulating foam blowing agents
- Thermal conductivity
- Chemical compatibility with foam matrix and adjoining product

* Responsible Care® is a program started by the American Chemical Council. Members continually improve their health, safety, and environmental performance; listen and respond to public concerns; assist each other in achieving optimum performance; and report their goals and progress to the public. See www.americanchemistry.com.

† TEWI includes both direct and indirect climate impacts of product manufacture and operation, but does not include calculations of the climate impact of manufacturing the product inputs and the climate impact at product disposal. LCCP is totally comprehensive over the full lifetime of the products through recycle of its components.

F-Center time-line

1986 ● Dr. Nobuo Ishikawa (F&F Research Center) and Masaaki Yamabe (Asahi Glass) participate in a US EPA Experts Panel on CFC Alternatives.

1989 ● Dr. Akira Sekiya organizes experts at the NIMC, even before funding is awarded.

1990 ● The Government of Japan begins funding research.

1993 ● Masaaki Yamabe wins EPA Stratospheric Protection Award.

1997 ● NIMC (Dr. Akira Sekiya), NEDO, and RITE win EPA Stratospheric Protection Award.

● Ozone Layer Protection Office (MITI, Japan) wins EPA Stratospheric Protection Award.

● Masaaki Yamabe wins EPA "Best-of-the-Best" Stratospheric Ozone Protection Award.

1998 ● The C5F-Team (NIMC and Nippon Zeon Company) wins EPA Stratospheric Protection Award.

6
Honda
Dream and it will happen[*]

Teams of engineers merging cutting-edge combustion, hybrid propulsion, advanced materials, aerodynamic, and regenerative technology to design Earth-friendly products that deliver value to the triple bottom line.

Soichiro Honda as a young man

Honda's environmental leadership was born out of the strong personality of founder Soichiro Honda, his business partner Takeo Fujisawa, and the unique culture that emerged from a small struggling company fascinated with technology and performance. Mr. Honda started his motorcycle company in 1948 after working for two decades as a mechanic and engine parts supplier. With an investment of only $3,200, he bought surplus engines and attached them to bicycles to provide affordable transportation. In just 12 years, this middle-school dropout built the world's largest motorcycle company.[1]

[*] The authors are grateful for interviews and supplementary assistance by the following Honda engineers and managers in Japan: Shigeru Fujii, Satoshi Fujitani, Hidekazu Kanou, Tomohiko Kawanabe, Ben Knight, Akito Kono, Mikio Kubo, Osamu Kuroiwa, Katsuyuki Morichika, Takao Nishida, Takaaki Nakai, Akihisa Nakamura, Kazuo Ochi, Hirotaka Ohki, Tsutomu Okuno, Takanori Shiina, Toshiyuki Suga, Kazushige Toshimitsu, Katsusuke Ueno, Hiroshi Umeno, Hideo Uzaki, Tadahiro Yaguchi, Kazuo Yamakawa, Ryutaro Yamazaki, Hiroshi Yamashita; and to American Honda managers David Raney and T. Michael Tebo.
[1] Readers who want to learn more about the history of Honda will want to start with Robert L. Shook's excellent book: *Honda, An American Success Story: Revolutionizing the Art of Management* (New York: Prentice Hall, 1988) and will also want to read *Dream: The Challenge of Creating and Progress*, published by the Honda History Project Group of Honda R&D in 1996. These books document extraordinary stories of leadership and risk taking that are so well known by employees that they were often told, with inventive variation, in our interviews.

Soichiro Honda was legendary for his quick temper when he demanded quality—he was sometimes called "Kaminari-san," "Mr. Thunder," after outbursts. He also was legendary for his ability to foster teamwork between workers and managers. He was a hands-on manager, rolling up his sleeves to solve problems at engineering centers and production plants. His business partner, Takeo Fujisawa, was the financial and marketing wizard who helped to create an organization where individuals could achieve their own ambitions by pursuing the goals of the team.

> " I consider making motorcycles to be my mission in life, and this, if nothing else, I want to do by creating an absolute beauty of form that is not inferior to what comes out of any other country. I always feel the desire to make this happen, no matter what. "
>
> *Soichiro Honda*
> Quoted in interviews with Honda

Japanese management encourages cooperation and consensus, perhaps because Japan's historic rice cultivation could only have succeeded with teamwork and mutual respect. Japanese culture, developed in a land with limited fossil fuel resources, depends on efficiency for sustainability. *Cooperation, consensus,* and *conservation* have important roots in Japan's respect for family values, art, culture, and nature. Soichiro Honda and Fujisawa Takeo declared in an early company principle: "Maintaining an international viewpoint, we are dedicated to supplying products of the highest performance."

Ironically, the 1950 Honda philosophy was five decades ahead of its time. "First, each individual should work for himself—that's important," said Mr. Honda. "People will not sacrifice themselves for the company. They come to work at the company to enjoy themselves. That feeling will lead to innovation." Honda welcomed new engineers to the company with a rousing speech telling them to "Think young and quest for the 'three joys': creating, buying, and selling."[2]

Build it and they will drive

Five decades later, teams of Honda engineers are still merging cutting-edge combustion, electric, hybrid, fuel cell, advanced materials, aerodynamic, and regenerative technologies to produce Earth-friendly products that deliver value to the triple bottom line—people, planet, and profits. Honda is the only global company that produces so many forms of powered equipment: motorcycle, automobile, marine, and power products.[3] Honda is also the world's largest engine-maker, producing more than 15.2 million annually.

2   Quoted in: Robert L. Shook, *Honda, An American Success Story: Revolutionizing the Art of Management* (New York: Prentice Hall, 1988).

3   Honda encourages the use of its engines in equipment made and marketed by other manufacturers. "Powered by Honda" means that a product has the fuel efficiency, low emissions, and reliability of Honda but other components made by cooperating companies. Honda engines power portable and standby electric generators, water pumps, air compressors, pressure washers, yard maintenance equipment, and farm equipment. Small Honda outboard marine engines are sold under several brands including Nissan.

At Honda, engineering ideas come from all product divisions, including many ideas with a big pay-off in environmental performance. For example, the Honda automobile engines that are among the cleanest in the world are reconfigured as outboard engines. The BF90/75 outboard motor is based on the Civic engine, the BF130/115 is based on the Accord engine, and the BF225/200/175 is based on the MDX engine. And Honda motorcycles, already the cleanest in the world, will further increase fuel efficiency. Honda engines provide quiet, reliable performance with lower fuel use and emissions. The cleanest Honda gasoline and natural gas prototype vehicles can reduce controlled emissions to almost unmeasurably low levels, and additional improvements will arrive soon.[4] This technical synergy of Honda engineers will be even more important as governments address problems of greenhouse gas emissions.

In Japan, companies are considered the "environmental top runner" when they set a new standard of excellence and stay well ahead of the competition in achieving the lowest emissions and highest fuel efficiency. Honda is recognized as the leader of almost every product category that it sells, almost every year.

President and CEO Hiroyuki Yoshino says,

> Through self-innovation, we are challenging ourselves to make the power train of today and tomorrow cleaner and more efficient. This will give new meaning to the words "Powered by Honda"—a phrase so important to our past and one that, I believe, will have even more power in the future.[5]

That future power is currently epitomized by the 2003 Honda Civic Hybrid, one of the most fuel-efficient five-passenger sedans ever sold in the world, which is certified as a Super Ultra-Low-Emission Vehicle (SULEV). The Hybrid Civic achieves 48 mpg (20 km per liter) (city) and 47 mpg (19 km per liter) (highway)—a 24–55% city, highway, and combined mileage improvement compared to other Honda Civics—the previous "top runners" (see Table 6.1).

4   The Honda ZLEV is not available yet.
5   Quoted in interviews with Honda.

	03M Civic Hybrid*	03M Civic LX AT	Percentage improvement
EPA city fuel economy (m/g)	48	31	55%
EPA highway fuel economy (m/g)	47	38	24%
Combined (m/g)	48	34	41%

* The Civic Hybrid SULEV/Zero Evap engine upgrade has comparable mileage with even lower emissions.

Table 6.1 Comparison of Civic sedans with automatic transmissions

Daniel Becker, Director of the Global Warming and Energy Program at the Sierra Club, says, "With Honda's decision to add hybrid models to their line-up of ULEV sedans, Honda has pulled into the lead of car-makers making greener products."[6]

The Honda Civic Hybrid achieves more than 30% improvement in combined city and highway fuel efficiency in a standard Civic chassis with no compromise in performance, safety, or convenience (due to a computerized operating system that requires no special attention from the drivers). The only driving difference is the noticeable silence when the "idle stop" feature turns the engine off at intersections and then restarts the engine with the depression of the gas pedal. The most conspicuous differences between the hybrid and standard Honda Civic will be the time and convenience of fewer stops for refueling and perhaps the attention owners get from friends and strangers who notice their cars and want to know more about them.

"It is remarkably satisfying to be driving down the road at 65-plus miles [104 km] per hour in this very quiet car knowing that it is getting at least twice the miles per gallon of most of the cars that surround you, with no sacrifice in comfort or performance," say Lucy Hull and Bart Chapin.[7] "I'd claim it's better than a dog for making friends," says Nancy Drucker.[8] "The women think it's cute . . . the guys say, 'Yeah, man, but has it got any acceleration?' I say, 'It's not a Ferrari. If you have to press that accelerator to feel good about yourself, go somewhere else,' " says Brian Kent.[9]

The Honda Civic Hybrid is the newest member of a family of top-runner products. Honda has an environmental top runner in every product category it sells: cars, motorcycles, scooters, marine engines, electric generators, farm equipment, lawnmowers, and more.

Although reducing emissions from automobiles is a global priority, motorcycles are emerging as the next generation of affordable and maneuverable transportation in developing countries—and prompt action now can prevent pollution and

6   Interview with Daniel Becker.
7   Quoted in: Natural Resources Council of Maine, *Hey, Cool Car! Hybrid Electric–Gasoline Vehicles in Maine* (Natural Resources Council of Maine): 3.
8   *Ibid.*: 8.
9   Jeffrey Ball, "For Mileage Fanatics, It's a Real Handicap to Have a Lead Foot," *Wall Street Journal*, June 5, 2002: front page.

actually increase the affordability of transportation in those countries. In 1999, 20 million motorcycles were sold worldwide, with about 80% sold in developing countries, and with China accounting for 45% of the market and India for 16% (see Figure 6.1). Honda is striving to produce "green" motorcycles that will provide mobility with the lowest possible environmental impact. Honda achieves high fuel efficiency and low emissions with the same strategy used for cars: lightweight material, fuel-injected 4-cycle and low-friction engines, and idle stop.

Honda can produce a motorcycle engine with half the pollution, and still improve the fuel efficiency of the typical product sold worldwide (see Figure 6.2). The savings in fuel with the Honda motorcycle actually pays for any increase in motorcycle price, and the owner saves even more with lower maintenance and higher resale value. For example, Honda has already achieved a 30% improvement in fuel efficiency by applying innovative environmental technologies developed for automobiles to the Wave-125 motorcycle. This was achieved by improving low- and mid-range torque, reducing engine friction with roller rocker arms and offset crankshaft, and computerizing ignition-timing control. A Honda prototype motorcycle, fuel-injected with an emission-control system, has emissions just $1/12$–$1/20$ of the "Euro 2" limits.

Larger motorcycles have achieved 20% higher fuel efficiency and low emissions with variable valve timing and lift electronic control (VTEC) engines incorporating

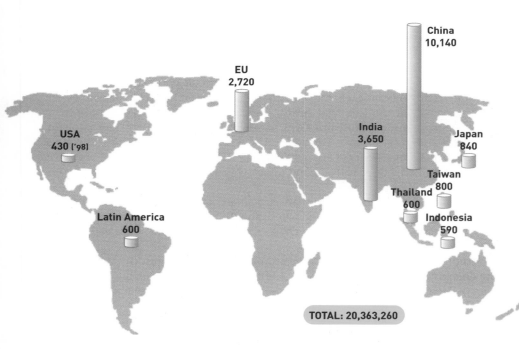

Figure 6.1 Annual motorcycle sales worldwide (1999) (× 1,000 units)

Source: Data from Honda

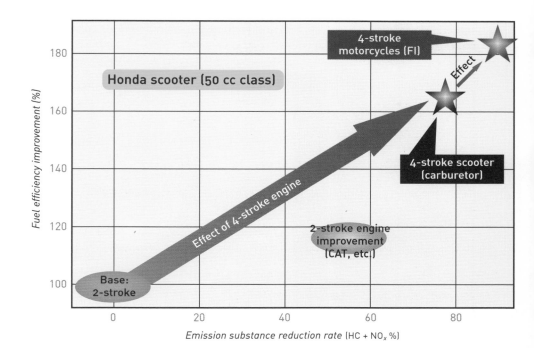

Figure 6.2 Emissions reductions from Honda fuel-injected (FI) small motorcycles

computer-controlled fuel injection, secondary air induction, and close-coupled three-way catalysts with dual oxygen sensors. Next-generation technology developed for motor sports will also be applied to basic motorcycles, including unicams, titanium valves, two-ring ultra-short pistons, and lightweighting and miniaturization.

Because Honda has a policy of producing its motorcycles in the countries where they are sold, a developing country can earn the same foreign exchange and create the same number of jobs that would be possible when producing inferior motorcycles. In addition, developing countries that choose to build environmentally superior technology will help protect the climate and will gain immediate health, economic, and safety benefits.

Honda satisfies the most stringent national environmental performance criteria in the markets where a vehicle model will be sold or manufactured; then the company supplies that environmentally superior vehicle to all of the markets where the model will be sold, including countries that do not require low emissions. This type of Honda leadership raises the bar on environmental competition, forcing competitors to try harder.

"Environmental leadership companies like Honda sometimes do more for the global environment than cumbersome regulations," says Yuichi Fujimoto of the JICOP. "In 1994, I served on a team of United Nations experts who discovered that

a few German, Japanese, and Korean companies were still selling CFC-12 air conditioners in their cars assembled in Vietnam. Honda and three dozen other multinational companies solved the problem by pledging to use the same ozone-safe technology in Vietnam as they do at home."

The youthful courage to succeed

Even as a young company, Honda faced crises with undaunted courage and leadership. In 1954 Mr. Honda responded to the threat of bankruptcy with the public announcement of its ambition to win "tourist trophy" (TT) motorcycle racing competitions. TT races are held on closed public highways; the best known is probably the Isle of Man race, which has been held since 1907. At first, this bold statement shocked employees, but they were soon motivated to succeed. Publicity for racing revived sales and brought profits that

Helen Tope (EPA Victoria), Sally Rand (US EPA) and Yuichi Fujimoto (JICOP) with the green Honda Insight

were shared by all. Honda formalized the lessons of translating racing success to sales with a four-part code: (1) face ambitious targets with a challenging spirit; (2) pursue original technology; (3) make efficient use of time; and (4) have self-confidence and joy in inventing for performance.

"Thinking young" became a theme in all aspects of Honda operations—openness, optimism, energy, and excitement. Thinking young meant that Honda would also seek excellence in advertising from any source. The campaign theme ultimately immortalized by The Beach Boys, "You meet the nicest people on a Honda," was conceived by University of California at Los Angeles (UCLA) students as a class assignment and marketed to Honda by their professor.[10]

Both Soichiro Honda and Takeo Fujisawa recognized the importance of announcing impossible technical goals and backing those ambitions with highly financed research programs. Shortly after production of the company's first automobile, Mr. Honda announced a plan to dominate racing and to win the famed Le Mans Formula One race. He knew that he could attract exceptional engineers by telling the world that Honda would be the best and fastest in the world.

This set a precedent for Honda's subsequent public announcements of seemingly impossible plans, such as Honda's goal to help protect the planet. The

10   Robert L. Shook, *Honda, An American Success Story: Revolutionizing the Art of Management* (New York: Prentice Hall, 1988): 34.

Figure 6.3 Honda Formula One racecar

obsession with racing reflected the ambitions of the company owners. Honda and Fujisawa used racing to attract and inspire engineers and to build teamwork and problem-solving. Racing involved clear results: win or lose; all or nothing; survival of the fittest. Winning engineers triumphed because they captured every technical advantage and extracted every last ounce of performance—where just two-hundredths of a second slower at the finish line could mean defeat. Winning required ambitious targets, passion, self-confidence, performance under pressure, courage, and stamina.

As a result of this process, Honda dominated class racing for many years. And, in 1965, the Honda team won its first Formula One victory at the Mexican Grand Prix, and a second in the 1966 Italian Formula One Grand Prix. In 1966, the Honda team became the first to capture all five solo World Motorcycle Championships (50 cc, 125 cc, 250 cc, 350 cc, and 500 cc classes). Also in 1966, Honda set a world record of 11 consecutive Formula Two victories. And in 1996 it also won the Indy/Championship Auto Racing Teams (CART) Series Championship.

Outside Honda, the racing program might have looked like advertising; but inside it was a carefully conceived strategy to push technology beyond its limits in ways that supported the goal of continuously improving consumer products. In 1968, Honda decided that his beloved automobile-racing program that had been devoted to power and speed could not produce an engine with low enough emissions to protect the planet. Characteristically, Soichiro Honda abruptly withdrew from Formula One competition, and publicly announced that Honda would now design the *cleanest engine* in the world.

The racing and business press could not believe what they were hearing. Some speculated that Honda was running out of money; others said it was quitting while it was ahead. Most claimed that the small Japanese company could not produce the low-emission engines that Detroit giants had announced were technically impossible. "Engineers were naturally disappointed when Honda withdrew from Formula One racing, but it wasn't long before we realized that we would be racing to protect the Earth for future generations. *And isn't that the ultimate race?*" remarked Hiroyuki Yoshino, President and CEO.[11]

11 Quoted in interviews with Honda

Winning the ultimate race

The best racing and research engineers at Honda devoted their full attention to a clean engine that would also achieve the goal of superior fuel economy. It was decided that the new engine would have the same power, equal or better fuel efficiency, and would operate on low-cost, globally available gasoline. Separate research teams competed and cooperated on developments. With each success, the teams were reorganized to pursue the most promising ideas, with experts forming the critical mass of necessary talent.

Honda was aware of the work by others on what became known as the "stratified-charge" engine, particularly the work by Russian L.A. Gussak on three-valve engines. This work demonstrated desirable fuel efficiency but with much higher emissions. "But, at least theoretically, we saw in the stratified-charge engine not only fuel economy, but the means for practical reliable emission control," says Tasuku Date of Honda Research and Development. "We then set out to achieve that controlled combustion, to convert a theoretical hope to a practical reality."[12]

A team of just eight engineers developed the revolutionary breakthrough technology, the Compound Vortex Controlled Combustion (CVCC) engine. The team included Tadashi Kume, who would later become CEO of Honda. The revolutionary CVCC was proven within a year, verifying Kume's genius.[13]

The original CVCC engine adds a small auxiliary combustion chamber located in the cylinder head (Figure 6.4). A rich fuel mixture is supplied to this auxiliary combustion chamber through its own intake passage and carburetor, and an extremely lean mixture is supplied to the main combustion chamber through its own intake passage and carburetor. The rich mixture is ignited by a spark plug in the auxiliary chamber, and the flame spreads through a torch opening to the lean mixture in the main chamber, ensuring reliable and complete combustion.[14] The geometry and timing is optimized to achieve a balance of power, fuel economy, emissions, durability, and reliability.[15]

The result was a stable, slow burn with a peak temperature low enough to minimize the formation of oxides of nitrogen, and a mean temperature held high enough and long enough to reduce carbon monoxide and hydrocarbon emissions.

---

12   T. Date, S. Yagi, A. Ishizuya, and I. Fujii, *Research and Development of the Honda* CVCC *Engine* (SAE Paper 740605; Washington, DC: Society of Automotive Engineers, February 1974): 1.

13   Robert L. Shook, *Honda, An American Success Story: Revolutionizing the Art of Management* (New York: Prentice Hall, 1988): 23.

14   A similar stratified-charge system with fuel injection had been developed in Germany but had not achieved the low emissions possible with the Honda design. See W.R. Brandstetter, G. Decker, H.J. Schafer, and D. Steinke, *The Volkswagen* PCI *Stratified Charge Concept-Results from 1.6-Liter Air Cooled Engine* (SAE Paper 741173; Washington, DC: Society of Automotive Engineers, February 1974).

15   T. Date, S. Yagi, A. Ishizuya, and I. Fujii, *Research and Development of the Honda* CVCC *Engine* (SAE Paper 740605; Washington, DC: Society of Automotive Engineers, February 1974); and S. Yagi, T. Date, and K. Inoue, NO_x *Emission and Fuel Economy of the Honda* CVCC *Engine* (SAE Paper 741158; Washington, DC: Society of Automotive Engineers, February 1974).

Figure 6.4 Latest valve mechanism showing auxiliary combustion chamber

Unlike emission controls based on chemical catalysts that deteriorate with use, the CVCC was designed to maintain low emissions for the life of the engine.

In 1973, Honda introduced the CVCC to the Honda Civic, the only car to satisfy the stringent 1975 emission standards under the 1970 US Clean Air Act and the 1975 Japanese standards.[16] It was the beginning of three decades of environmental progress leading to the current engines and hybrid systems that have almost

16 US and Japanese Standards 1975:

	CO	HC	NO$_x$
Japanese Standard (g/km)	210	0.25	120
Original Muskie Standard (g/km)	211	0.25	193

immeasurable emissions. Honda achieved the US standard—officially certified by the Environmental Protection Agency—with regular gas and with no catalytic converter. Grasping the significance, publications as diverse as *Reader's Digest* and the *Society of Automotive Engineers Journal* praised the achievement. The December 1975 *Reader's Digest* article "From Japan—A 'Clean Car' That Saves Gas" embarrassed Detroit with its strong words:

> Then came another surprise. Honda engineers installed their stratified-charge cylinder head and carburetor on the eight-cylinder engine of a 1973 Chevrolet Impala. When the Environmental Protection Agency tested this big-engine, full-size car, it found that Honda's modifications had reduced emissions substantially below the 1975 federal limits and that fuel economy was increased.[17]

Honda estimated that the new clean engine added about $170 to the cost of a conventional engine. This cost would be offset by fuel savings; and it avoided the typical $350 cost of catalytic converters, which decreased power, increased fuel costs, and would require more maintenance.

"The CVCC helped Honda earn its reputation for quality and engineering excellence. It is ironic that Detroit's claim that no one could meet that standard helped the public appreciate what had been accomplished," said Aki Nakamura, Technical Vice President (retired). Customers embraced the new Honda cars equipped with the 1,500 cc CVCC engine and purchased 800,000 by 1980.

Back to the finish line

Having achieved its clean engine goal, Honda returned to automobile racing—and at the same time expanded its race program to compete in environmental competitions such as the "World Solar Challenge," the world's top solar car race (Figure 6.5). "Racing solar cars is very competitive and pushes every technology to the limit," says Hiroyuki Ozawa of Honda R&D.[18] The balance between electric energy generated from solar cells and energy consumption dictated by the aerodynamics and rolling resistance determines the cruising speed of a solar car. "Solar car racing taught us to strive for the full potential of the car, in addition to boosting the efficiency of all individual components," says Ozawa. In 1999, Honda-powered drivers finished first, second, and third in the CART World Series for the third

---

17 EPA tested three Honda-powered vehicles and the 350 cubic inch Chevy Impala, confirming that the CVCC technology reduced emissions for both large and small engines to levels within the 1975 EPA limits. See US EPA, *An Evaluation of Three Honda CVCC Powered Vehicles* (Washington, DC: US EPA, December 1972); and US EPA, *An Evaluation of a 350 CID CVCC Powered Chevrolet Impala* (Washington, DC: US EPA, October 1973).

18 Y. Shimizu, Yasuyuki Komatsu, Minoru Torii, and Masato Takamuro, "Solar Car Cruising Strategy and its Supporting System," *Journal of the Society of Automotive Engineers Review* 19 (1998): 143-49; and H. Ozawa, S. Nishikawa, and D. Higashida, "Development of Aerodynamics for a Solar Race Car," *Journal of the Society of Automotive Engineers Review* 19 (1998): 343-49.

Figure 6.5 Honda solar racecar

HONDA INNOVATION AND INVENTION

- Soichiro Honda was awarded more than 470 patents.
- Honda engineering firsts have come in combustion, anti-lock brakes, four-wheel steering, electronic fuel injection, continuously variable transmissions, aluminum die casting, and forming.
- More than 300 patents are pending for innovation on the hybrid Insight and Civic.

consecutive year. In 2000, Honda returned to Formula One Grand Prix auto racing, the sport it had dominated in the 1980s.

Engineers were naturally disappointed when Honda withdrew from Formula One racing, but they soon realized his larger goal, that they were now racing to protect the Earth for future generations. They were now striving to win the ultimate race. For example, Honda marketed electric cars and vans under very favorable leasing terms that included complete maintenance and insurance, but was still unable to satisfy customers. It then switched to the hybrid car, because it could achieve comparable environmental performance as well as satisfy consumer demands for vehicle range, comfort, and simplicity of operation.

Honda's top managers and CEOs have always been engineers. "It is not that we ever decided to have a president with an engineering background, but an underlying concept of this firm is that rather than being driven to realize profits, we want to make superior engines. When a company thinks in these terms, its customers will be satisfied, and the profits will eventually be there too," says Tadashi Kume.[19]

Honda products have been first to achieve the vast majority of the most daunting European Union, Japanese, and North American emissions standards; and Honda offers these clean and fuel-efficient products in every market—even markets without stringent environmental requirements.

This environmental superiority is achieved at Honda as a result of the strong ethic of innovation and inspiration developed by its founders, and by setting

---

19 Quoted in: Robert L. Shook, *Honda, An American Success Story: Revolutionizing the Art of Management* (New York: Prentice Hall, 1988): 189.

challenging goals and financing research and development heavily. This "ethic" has created a synergy of imagination and technical know-how between engineers from all its product divisions. Environmental excellence from each division is shared company-wide, and the best environmental performance engineers are rotated throughout the racing programs to reward and inspire even more extraordinary inventions.

The synergy of engineering

Honda encourages "synergy of engineering" between product areas through its centralized research centers with "crossover meetings" of engineers, and by encouraging every employee (called "associates" at Honda) to make suggestions about the company. Consider how the company's patented variable valve timing and lift electronic control (VTEC) evolved over time. VTEC engines essentially take one intake valve per cylinder out of operation at low engine speeds to produce a precise and very lean air–fuel ratio. Operating with just one intake valve induces a high degree of swirl in the combustion chamber, leading to faster and more complete combustion. At higher engine speeds, both intake valves operate to increase engine-breathing capacity. This results in double the intake-valve area, achieving higher power output.[20]

VTEC was first developed for racing motorcycles to achieve higher power at every engine speed, and to allow operation at very high revolutions per minute (rpm). After VTEC had proven itself at the races, it was first commercialized in the 1983 CBR400F motorcycle and then withdrawn when other engine designs provided needed performance. In 1989, VTEC was introduced in the Acura NSX sports car and Integra, in configurations designed to deliver high-power performance. Later it was reapplied to motorcycles as HYPER VTEC (CB400SF: 1999; VFR800FI: 2002YM).

> In the early eighties, our motorcycle racing team was desperate to wring out every bit of horsepower. The intricate mechanisms of VTEC were less daunting to motorcycle engineers because everything on a motorcycle is miniaturized to fit between the rider's legs. The demands of racing inspired the engineers to push the technology to its limits, and in the end that won road races—*and is helping win the race for sustainable development*. At that time, no one could have imagined that variable valve timing could reduce emissions, and no environmental regulatory authority was demanding the low emission levels that are now proven to be necessary for air quality and climate protection.[21]

Honda perfected VTEC to provide faster and more complete combustion, resulting in very high fuel efficiency and reduced hydrocarbon and carbon monoxide

---

20   K. Horie, K. Nishizawa, T. Ogawa, S. Akazaki, and K. Miura, *The Development of a High Fuel Economy and High Performance Four-Valve Lean Burn Engine* (SAE Paper 920455; Washington, DC: Society of Automotive Engineers, February 1992).
21   Personal interview with Osamu Kuroiwa, Deputy General Manager of Environment and Safety Planning Office, Honda, February 2002.

emissions necessary to satisfy increasingly stringent California and global emission standards. Honda engineers were among the first to realize that the people of California would demand increasingly stringent standards, that national standards would quickly match any more-stringent state standards, and that vehicle manufacturers would want the manufacturing and marketing convenience of supplying the same car to all states.

The holistic "grail" of efficiency

The fuel-efficient version of the VTEC engine, called the VTEC-E ("E" for economy), was introduced in the 1992 Honda Civic HX Coupe and soon became the foundation for many of Honda's breakthroughs in environmental technology. Honda added three-stage, four-valve-per-cylinder timing, incorporating an engine management system that can assure highly accurate air–fuel ratios controlled by an air–fuel ratio sensor using zirconia (ZrO_2) for the 1995 Civic and Acura.[22]

Traditionally, engineering goals such as fuel economy, low emissions, performance, and safety were viewed as mutually exclusive. For example, early engineering approaches to safety often made cars heavier, which compromised both performance and fuel economy. "The old thinking was that power would always be sacrificed for fuel efficiency and even that fuel efficiency would be sacrificed to reduce emissions. This was true when engine designers just bolted on exhaust gas recirculation pumps, reduced compression ratios with shorter piston stroke, or made other crude compromises."[23] Honda engineers made it their goal to demonstrate unequivocally that fuel efficiency and engine performance could be achieved simultaneously. "We went back to the basics of physics and chemistry to learn that there were no limits. We embraced complex mechanisms and controls, and we experimented bravely until we succeeded. Today it is obvious that the goal is to turn all the fuel into power, with no toxic or greenhouse gas emissions."[24]

In 1998, Honda's revolutionary designs achieved the first gasoline engine certified by California and the US EPA as meeting the Ultra-Low-Emission Vehicle (ULEV) standard—which was previously deemed impossible. In the same year, Honda introduced its Natural Gas Vehicle (NGV), achieving even lower emissions— "almost zero"—for most pollutants and a 20% reduction in greenhouse gas emissions. Remarkably, Honda achieved ULEV- and NGV-level emissions while actually improving engine output and fuel efficiency—proof again that environmental protection can be "no compromise."[25]

22   *Ibid.*
23   Kazushige Toshimitsu, Senior Chief Engineer, Advanced Product Planning Division, interview, February 2002.
24   Hidekazu Kanou, Chief Engineer, Advanced Product Planning Division, interview, February 2002.
25   N. Kishi, S. Kikuchi, Y. Seki, A. Kato, and K. Fujimori, *Development of the High Performance L4 Engine ULEV System* (SAE Paper 980415; Washington, DC: Society of Automotive Engineers, February 23, 1998).

In 1999, Honda announced the ultimate combustion engine—the Zero-Level Emission Vehicle (ZLEV) based on the 2.3 liter, 4 cylinder engine found in 1998 and later Honda Accords. This vehicle has just one-tenth of the emissions of California's ULEV standard, the most stringent standard in the world both then and now (see Table 6.2). The name "ZLEV" is a term created by Honda. Honda defines the ZLEV as achieving one-tenth the ULEV standard, with nearly immeasurable emissions that are lower than the air the vehicle drives in ("negative emissions effect").[26]

	Non-methane organic gas (NMOG)	Carbon monoxide (CO)	Nitrogen oxides (NO$_x$)
Tier 0 Standard	0.41 total hydrocarbons (THC)	3.4	1.0
Tier 1 Standard	0.25 non-methane hydrocarbon (NMHC)	3.4	0.4
LEV Standard	0.075 @ 50k miles 0.90 @ 120k miles	3.4 @ 50k miles 4.2 @ 120k miles	0.05 @ 50k miles 0.07 @ 120k miles
ULEV Standard	0.040 @ 50k miles 0.055 @ 120k miles	1.7 @ 50k miles 2.1 @ 120k miles	0.05 @ 50k miles 0.07 @ 120k miles
SULEV Standard	0.010 @ 120k miles	1.0 @ 120k miles	0.02 @120k miles
01 Civic GX (NGV) test results	0.0084 @ 120k miles	0.1 @ 150k miles	0.01 @ 150k miles
Honda SULEV test results	0.0084 @ 120k miles	0.0155 @ 120k miles	0.0161 @ 120k miles
Honda ZLEV test results	< 0.004 (w/100k-mile aged catalyst)	< 0.17 (w/100k-mile aged catalyst)	< 0.02 (w/100k-mile aged catalyst)

Table 6.2 California LEV II Standards and Honda SULEV and ZLEV test results

The "next-generation" 2 liter, 4 cylinder Honda engines, which have debuted in Japan and the USA and will be offered worldwide by 2005, achieve ULEV emissions and a 10–20% boost in fuel efficiency as well as a reduction in greenhouse gas emissions plus higher performance including low-speed torque. Honda ZLEV and NGV vehicles achieve emission levels that are almost the same per kilometer as emissions produced by a clean mix of the power plants necessary to power electric vehicles (see Figure 6.6).

"A car equipped with this engine could drive through a high smog area, and the smog-producing emissions coming out of the tailpipe would actually be lower than in the surrounding air," said Nobuhiko Kawamoto, Honda's former President and CEO. "This engine represents a feasible approach, one we believe can be

26   N. Kishi, S, Kikuchi, N. Suzuki, and T. Hayashi, *Technology for Reducing Exhaust Gas Emissions in Zero Level Emission Vehicles (ZLEV)* (1999-01-0772; Washington, DC: Society of Automotive Engineers, January 3, 1999).

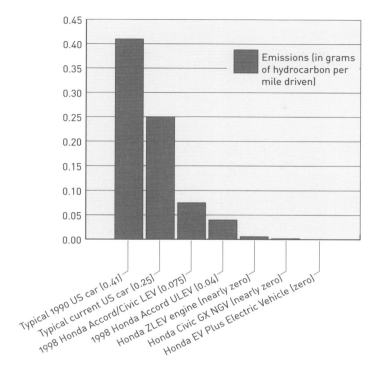

Figure 6.6 Clean, cleaner, cleanest engine

applied to Honda products."[27] The Honda ZLEV does depend, however, on low-sulfur reformulated fuel currently available only in California, Japan, and some European countries.

At Honda, the future is now

"Today everyone is excited about the fuel cell cars that will dominate markets after 2010, but we have not forgotten the importance of making next year's cars more friendly to the environment. We will continue to use synergy between our projects," says Tomohiko Kawanabe, Managing Director of Honda R&D. "For example, our hydrogen fuel cell vehicle uses the electric motor drive system developed for the EV Plus electric car, the energy management systems used in the Insight and Civic gasoline–electric hybrid car, and the high-pressure gas storage tank technologies developed for the natural gas-powered Civic GX."

27   Quoted in interviews with Honda.

OUTRUNNING THE VOCABULARY OF REDUCED VEHICLE EMISSIONS

Emission standards can be defined by scope (the list of chemicals controlled), stringency (the grams per mile allowed), and life-cycle (as new and at designated vehicle mileage). Until 2002, the standards covered toxic chemicals and chemical precursors to smog but not carbon dioxide or greenhouse gases. The US EPA sets national vehicle emission standards, but in some circumstances allows states to set more stringent standards. California has historically been the only state to exercise this right consistently. These standards are expressed in emissions per mile and, beginning in the 1990s, were translated by EPA into acronyms for convenience.

At first, it was easy to describe improvements, but now regulators are reaching a vocabulary dead end, and California is defining its own acronyms. The first EPA name was "Low-Emission Vehicle" (LEV). The next was logically named "Ultra-Low-Emission Vehicle" (ULEV), followed by the "Super-Ultra-Low-Emission vehicle" (SULEV) and then a designation of "Inherently Low-Emission vehicles" (ILEV). Sensing that adverbs were running out, the EPA just recently defined the "National Low-Emission Vehicle" (NLEV) with subcategories of "tiers" and "bins." California chose instead the designation of "Zero-Emission Vehicle" (ZEV). For example, the California ZEV is currently equivalent to the EPA's Tier 2 Bin 1 NLEV. In July 2002, California announced plans to control carbon dioxide from cars—no doubt starting a new round of cleaner-vehicle acronyms.

In 2001, the Civic featured a safety chassis and a new power train incorporating a further modified 1.5 liter VTEC lean-burn engine, with a continuously variable transmission. Fuel economy was improved by 10% in North American markets and by an additional 8% in Japan and other countries that have embraced low-sulfur fuel that allows direct injection.[28] Emissions were reduced dramatically to satisfy the most stringent standard in the world—the US ULEV standard.

Relentless pursuit of
life-cycle climate performance (LCCP)

Since 1966, Honda has used LCCP to measure the total CO_2 emissions from every aspect of vehicle material supply, manufacturing, use, and disposal/recycling. A typical vehicle produces 80% of CO_2 emissions from fuel use; 18% from materials, manufacturing, and transport; and 4% in maintenance and disposal. Honda is working to reduce all of these emissions by increasing the use of recycled materials, by eliminating unnecessary weight and content, by making manufacturing more efficient, and by transporting parts and vehicles by the most energy-efficient option. The highest priority is to reduce the 80% of emissions from fuel consumption.

A conventional vehicle must have an engine large enough to provide acceptable power under the most demanding circumstances—fully loaded, high altitude,

---

28   M. Matsuki, H. Ohura, K. Watanabe, and Y. Kinoshita, "Development of a 1.5 l Lean-burn Engine for the 2001 Civic," *Honda R&D Technical Review* 12.2 (October 2000).

Figure 6.7 Honda hybrid engine, motor drive and automatic transmission

steep hills, and strong acceleration. During braking, a conventional vehicle wastes the kinetic energy of motion by converting it to heat energy via the brakes. Fuel is also wasted when engines idle. A hybrid takes advantage of the best traits of each power system—the clean nature of electricity and the high performance, range, and infrastructure of internal combustion engines.

Honda places its low-emission, clean-burn gasoline engines in hybrid electric–gas Insight and Civic models (Figure 6.7). Hybrid propulsion essentially recycles energy that would otherwise be lost during braking, and uses that power for acceleration at a later time. During acceleration and other times of heavy engine load, the electric motor assists the gasoline engine by providing additional power, resulting in improved acceleration without compromising fuel economy. This allows a significant reduction in engine displacement and higher engine efficiency.

At cruising speeds when the engine load is lower, the computerized motor assist system lets the ultra-efficient gasoline engine maintain the vehicle's speed. During deceleration, the electric motor becomes a generator and converts the energy into electricity. When the car comes to a temporary stop, at a traffic light for example, the engine shuts off automatically. Then it restarts immediately when the driver lets off the brake pedal and depresses the gas pedal on automatic transmission vehicles, or depresses the clutch and puts the car into first gear on manual transmission vehicles.[29] The Insight Hybrid is a two-passenger car built from the ground up for fuel efficiency and low emissions. The Civic Hybrid is a five-passenger, four-door sedan.

The Insight achieves its top-runner fuel economy through three technical break-throughs. The integrated motor assist (IMA) system (with idle stop and regenerative breaking) contributes 30% to fuel efficiency; the advanced engine contributes 35%; and the lightweight aluminum body, low rolling-resistance tires, and aero-dynamic body design contribute 35%. The integrated electric motor system is sized precisely to operate without surplus or deficit, and to minimize the size and weight of batteries to improve fuel economy without sacrificing vehicle performance.

The hybrid propulsion system achieves a 94% fuel economy advantage with no sacri-fices in dynamic performance when tested on equivalent 500 kg vehicles.[30] The aluminum Insight body is 40–50% lighter than a com-parably sized steel body using a combination of both Monique and space frame structure.[31] The use of aluminum also reduces the total greenhouse gas emissions from the manufac-turing of the vehicle, with the highest reduc-tions achieved with recycled aluminum. Chapter 1 of this book, on Alcoa, elaborates the importance of aluminum to climate protection. Here Honda's experience building the Acura NSX sports car paid off, because Honda engineers were able to apply new con-struction techniques, such as the use of extrusions and castings in the body.

The Insight is the world's most aero-dynamic mass-produced automobile, with a

> "Hybrid technology is ready to transform markets, demonstrating an extraordinary 30% improvement in fuel efficiency while maintaining the acceleration and speed of conventional Civics," says Honda. "The Honda Civic is the best-selling compact car in America and one of the pillars of the Honda brand worldwide. To add hybrid power to the Civic line-up is an example of the faith and confidence we have in our hybrid technology."* Hybrid vehicles are currently available only from Honda and Toyota, but DaimlerChrysler, Ford, and General Motors (GM) have announced plans to introduce hybrid vehicles: Ford Escape SUV (sport utility vehicle) in December 2003; GM in large pickup trucks in 2004; and DaimlerChrysler in future SUVs.
>
> * Source: Honda press releases.

---

29 K. Nakano, K. Aoki, B. Knight, S. Kajiwara, H. Sato, and Y. Yamamoto, *An Integrated Motor Assist Hybrid System* (2216; Washington, DC: Society of Automotive Engineers, 2001); and S. Ochiai, K. Uchibori, K. Hara, T. Tsurumi, and M. Suzuki, "Development of a Motor Assist System for a Hybrid Car—Insight," *Honda R&D Technical Review* 12.1 (April 2000): 7-14.

30 Y. Hasegawa, S. Aoyagi, T. Yonekura, and H. Abe, "Energy Efficiency Improvement for a Series Hybrid Vehicle," *Honda R&D Technical Review* 12.1 (April 2000): 79-84.

31 A Monique frame structure replaces traditional rigid frame rails by engineering the sheet metal to provided the necessary strength. The space frame structure is a further refine-ment of the Monique structure that collapses in a collision in a manner that protects the occupant areas.

0.25 coefficient of drag and a weight of only 1,856 lb (841 kg). The aluminum body achieves high standards of safety, performance, stiffness, and cost savings relative to previous aluminum bodies. For example, to minimize head-on collision damage, the front and rear frame parts are made of extruded aluminum with different cross-sections in such a way that the front part will be compressed and collapsed and the rear part will bend and alleviate any penetration and impact on the body. The energy absorption rate is improved and the weight is reduced significantly. Lower weight improves fuel efficiency and improves braking and handling—possibly avoiding collisions that would occur with heavier steel bodies. The Insight aluminum body meets all US, European, and Japanese safety standards, including the 2003 safety standards for side impact and head injury protection.[32]

> If all cars in the Honda Civic size class were replaced by a Honda Civic Hybrid, 448 million gallons (1,702 million l) of gasoline would be conserved, keeping 4 MMt CO_2 out of the air. This would reduce atmospheric CO_2 levels by the same amount as a 1.2 million-acre (485,000 ha) forest, roughly the size of Glacier National Park, Montana.

The 2003 ULEV Hybrid Civic achieves top-runner fuel economy with a 30% mileage improvement compared to the 2003 Civic with only a gasoline engine, but in some respects is the same as the traditional Civic.[33] It sells for about $18,000—only $3,000 more than a comparably equipped standard Civic. The final cost of the hybrid feature is reduced by credits available in Europe, Japan, and the USA and offset by fuel savings. About 95% of that extraordinary improvement is the result of an integrated gasoline engine/electric motor system, with about 5% coming from inconspicuous improvements in aerodynamics (front air dam and rear spoiler, revised underbody panels) and highly efficient power steering and special wheels and tires to reduce rolling resistance and inertia.

The engine in the Civic Hybrid has several breakthrough technologies including dual and sequential ignition, and VTEC variable-valve control with cylinder deactivation. Dual and sequential ignition features two separately controlled spark plugs per cylinder to maximize combustion. VTEC cylinder deactivation shuts down three of the four

AWARDS FOR HONDA HYBRID CARS

- *Popular Mechanics* Design and Engineering Award
- *Automobile Magazine* "2000 Technology of the Year"
- *Popular Science* Best of What's New Award
- *American Woman Motorscene* "Most Likely to Change the World"
- Clean Car Coalition "Clean Car Salute"
- Edmunds.com "Most Significant New Vehicle"
- Sierra Club Environmental Engineering Award
- US EPA 2000 Climate Protection Award
- US EPA "Most Fuel-Efficient Car in America" 2000, 2001, 2002, and 2003
- American Council for an Energy Efficient Economy (ACEEE) "Best Vehicle in Class"

---

32 M. Saito, S. Iwatsuki, K. Yasunaga, and K. Andoh, "Development of Aluminum Body for the Most Fuel Efficient Vehicle," *Japan Society of Automotive Engineers Review* 21 (2000): 511-16.
33 The Civic Hybrid has improved aerodynamics around the wheel wells, alloy wheels as a standard feature, and leaves out the fold-down rear seat to accommodate the batteries.

cylinders during deceleration to significantly increase the amount of electrical energy that can be recovered.

"Think of this integrated motor assist (IMA) system as an electric supercharger that provides additional performance without using much energy because it recovers the Civic's momentum to recharge itself," suggests Csaba Csere of *Car and Driver Magazine*.[34]

In 2002, Honda announced another revolutionary goal for environmental protection: to market a fuel-cell vehicle, operating on renewable energy, that achieves zero toxic emissions and zero greenhouse gas emissions from its propulsion system.[35] The company also intends to achieve two or three times the energy conversion efficiency of the most efficient gasoline engine and to attract customers with a vehicle that is easy to use, has the same dynamic performance, safety, durability and reliability of today's passenger cars. "Of course this is daunting and almost impossible," said Honda Research and Development Managing Director Tomohiko Kawanabe, "But it is clearly necessary for future generations. It is our duty and our challenge."

Looking back, it is possible to define some Honda environmental technologies as *evolutionary* and others as *revolutionary* as shown in Table 6.3. Revolutionary technology is entirely new and represents a breakthrough in environmental performance. Evolutionary technology fine-tunes existing technology to further increase power and fuel efficiency and to reduce emissions.

Tomohiko Kawanabe expands on the evolutionary and revolutionary processes of technology innovation at Honda:

> Each breakthrough in environmental technology required self-confident engineers who were willing to dismiss engineering skepticism and tackle problems again and again as environmental regulations matured. Revolutionary breakthroughs like the CVCC were continuously improved until we reached the limits of emission reduction. When we couldn't go any further, we jumped to a new revolutionary technology and then continuously improved that. I am very proud that Honda combustion engines with hybrid drive can environmentally outperform current electric and fuel cell propulsion, but I will celebrate when new technology has even less impact on humans and nature.

Honda, like all vehicle manufacturers, faces daunting challenges in improving the environmental performance of vehicles while satisfying safety performance. Intuitively, most people believe that increased weight and size improves safety, but this is not true. Heavy vehicles protect the occupants against being crushed in an accident, but heavy vehicles take longer to stop, are less maneuverable at high speeds, may roll over more easily, and can cause tire failure leading to complete loss of control.

Furthermore, any advantage in size is offset in accidents with other large vehicles. Fortunately, Honda engineers are making breakthroughs in safety. Car bodies are now shaped and formed to be stronger without increasing weight.

---

34 Csaba Csere quoted in *Car and Driver Magazine*, May 2002.
35 Greenhouse gas emissions will occur from vehicle manufacturing, transport, and disposal, and from fuel refining.

Revolutionary improvements	Evolutionary improvements
Compound Vortex Controlled Combustion (CVCC) 1973	Branched circuit CVCC 1976 (longer torch passage oriented toward approximate center)
	Multi-opening torch 1979 (five-opening torch spreads flame all over the higher-compression main combustion chamber with exhaust gas catalyst for CO and HC)
Variable valve timing and lift electronic control system (VTEC) 1983 motorcycles, 1990 cars	1.5 liter lean-burn engine 1992 (for the Civic)
Variable valve timing and lift electronic control system (VTEC) 1983 motorcycles, 1990 cars	1.5 liter 3 stage VTEC combined with the continuously variable transmission (CVT) 1996 (4 valve, variable swirl, valve timing and lift managed by 3 stage control—inactive, medium-speed, and high-speed cams).
	1.8 liter VTEC lean-burn 1998 (for the Accord)
	Evolved VTEC lean-burn with CVT 2000
	Evolved 1.5 liter lean-burn 2001 (reduced engine friction, newly designed intake port and combustion chamber, direct injection three-way NO_x adsorption reduction type catalyst)
	2.0 liter lean-burn i-VTEC 2000 (resin intake manifold, improved cylinder head, optimized valve timing, friction reduction, advanced exhaust, quick catalyst heating)
Ultra-Low-Emission Vehicle (ULEV) VTEC	ULEV VTEC 1997 (adaptive "self-tuning" precise cylinder-specific combustion control, air-gap exhaust for rapid heating of new three-way catalyst with precious metal layering and increased surface area)
Super-Ultra-Low-Emission Vehicle (SULEV)	SULEV 2000 (ULEV VTEC technology with predictive fuel control)
Hybrid gas–electric	World's lightest 1.0 liter gasoline engine merged with ultra-thin electric motor and regenerative braking for improved efficiency and added power when needed, earning the best EPA mileage ratings in history.

Table 6.3 Honda's revolutionary and evolutionary environmental technologies

This is all part of Honda's effort to strive for both revolutionary and evolutionary product improvement. Given its history of successful innovation, it is clear that Honda's dream for better environmental products will continue to come true.

Honda time-line

1935 ● Soichiro Honda races at the All-Japan Speed Rally, setting a speed record before being injured seriously.

1948 ● Soichiro Honda founds his company with only ¥1 million (< $10,000) investment.

1949 ● Co-founder Takeo Fujisawa joins the company.

1954 ● Honda factory motorcycles enter their first race in São Paulo, Brazil.

1955 ● Honda wins All-Japan Motorcycle Endurance Road Race in 350 cc.

1958 ● Super Cub motorcycle released.

1961 ● Honda wins top five positions in both 125 cc and 250 cc classes at Isle of Man Race.

1964 ● Honda announces participation in Formula One car racing.
● The GB30 with 4-stroke engine was released in the marine engine category, which generates less noise and vibration than 2-stroke models, and consumes less fuel and produces cleaner emissions.

1965 ● Honda Formula One car wins the Mexico Grand Prix Race.

1966 ● Honda wins five classes in the World Motorcycle Championship Grand Prix Race.

1967 ● Production of automobiles started at Suzuka, Japan.

1970 ● US Clean Air Act and Japanese regulators set first stringent vehicle emission standards.
● Honda N600 car sold in the USA (air-cooled, chain drive).

1971 ● Honda invents Compound Vortex Controlled Combustion (CVCC) engine.

1972 ● Details of CVCC low-emission engine system announced; CVCC engine complies with the US Muskie Act of 1975.

1973 ● Honda achieves 1975 Japanese and US emissions standard without catalysts.
● Ford, Chrysler, Isuzu, and Toyota sign licensing agreement with Honda for CVCC technology.

1974 ● CVCC engine development group awarded "Society Prize" by the Society of Automotive Engineers.

1977 ● Civic CVCC ranked first in US EPA Fuel Economy Test.

1983 ● CVCC Chief Engineer Tadashi Kume becomes president, replacing retiring Kiyoshi Kawashima.
● VTEC first developed and implemented for CBR400F motorcycle (reapplied to motorcycles as HYPER VTEC in the 1999 CB400SF and the 2002 VFR800FI).

1985 ● Honda CRX-HF is the first mass-produced 4 cylinder car to break the 50 mpg (21 km per liter) barrier.

1986 ● Honda rated number one in J.D. Power Consumer Satisfaction Index.
   ● Honda captures Formula One Constructors' Championship.
1987 ● Acura rated number one and Honda number two in J.D. Power Consumer Satisfaction Index.
1990 ● VTEC engine first introduced on Acura NSX sports car, later to become the foundation for many of Honda's breakthroughs in environmental technology.
1992 ● Honda captures sixth consecutive Formula One Constructors' Championship.
   ● First outboard motor to comply fully with the European Lake Borden emission-control regulations.
1993 ● Honda wins the World Solar Challenge with the "Honda Dream" (held every three years).
1994 ● Electric scooter introduced.
1995 ● Civic becomes the first gasoline-powered vehicle to meet the California LEV standard; Honda achieves 2000 California standards requiring fleet average emissions at LEV—70% lower than the toughest federal standards for reducing emissions.
1996 ● Honda wins the World Solar Challenge.
1997 ● The BF130, a low-pollution marine engine, developed by Honda; all models are already well below the 2006 marine engine emission limits set by the US EPA.
   ● Honda Accord EX with automatic transmission first to achieve 2000 California standards for ULEV engines.
   ● Honda becomes first auto manufacturer to sell LEVs voluntarily in all states, with both Accord and Civic models.
   ● Honda EV Plus is first application of advanced nickel-metal hydride batteries in an electric vehicle.
   ● Honda announces a gasoline-powered internal combustion engine that is virtually pollution-free.
   ● Honda announces that it will gradually abandon manufacture of 2-cycle motorcycle engines and switch to more environmentally friendly 4-cycle engines.
1998 ● Honda distributes "Best-In-Class" ULEV Accord in California, Connecticut, Massachusetts, and New York—with "no-compromise" engineering that achieves the lowest emissions while boosting power by 15% and increasing fuel economy and decreasing emissions of greenhouse gases.
   ● Honda GX SULEV Accord is first SULEV sold in the USA and was certified as the cleanest internal combustion vehicle ever tested, with a 20% reduction in greenhouse gas emissions.
1999 ● Honda Insight is first gasoline–electric hybrid vehicle for sale in the US.
2000 ● Honda Accord Super-Ultra-Low-Emission Vehicle (SULEV) is first vehicle certified to meet California's 2004 standard and was the cleanest gasoline-powered vehicle tested.
   ● Honda Accord SULEV awarded the California South Coast Air Quality Management District's 2000 Clean Air Award for Advancement of Air Pollution Technology.

2000 • Honda distributes Civic and MDX vehicles meeting California ULEV standards to all US states.

2001 • Honda MY Civic is the first car to meet California's ULEV standard in all states.

• Honda improves Insight's fuel efficiency by 5% to 9% and also improves power, performance, and safety.

• Honda announces Hybrid snowblower HS1390I; the snowblower is equipped with a Honda e-SPEC engine, an environment-friendly powerplant that surpasses the latest EPA regulations; and fuel economy improves 10% against same-type snowblower.

2002 • Honda hydrogen-powered fuel-cell vehicle (FCX) first to be certified by the California Air Resources Board as a Zero-Emission Vehicle (ZEV) and by the US EPA as a Tier 2 Bin 1, National Low-Emission Vehicle (NLEV), the lowest national emission rating.

• All Honda and Acura models sold in US meet or exceed Low-Emission Vehicle (LEV) levels.

7

Seiko Epson
Some day all products will be made the Epson way *

EPSON

Epson dreams of a world without disposable batteries—a world with portable computers, cell phones, and personal entertainment systems powered by the kinetic energy of keystrokes, motion, and the sun. It is a world where "compact manufacturing" wastes no energy or natural resources. A world where the natural environment is preserved and every action of a company and its workers supports the economic vitality of their communities. Epson strives for a sustainable society based on mobile devices, reducing the emissions of greenhouse gases from power generation now necessary to charge batteries, and reducing the improper or expensive disposal of toxic battery waste.

Seiko Epson has its headquarters far from Tokyo, in Suwa, in the foothills of the "Japanese Alps." This privately owned company, the parent of a group of companies with 69,000 employees spanning 39 countries on six continents, has made a commitment to protect the global environment that far exceeds anything mandated by law, and which goes beyond what even the most green companies think is technically possible.

The founding companies that became Seiko Epson have been in Suwa since the early 1940s. From the start, they have shared the goal of preventing pollution and of living in harmony with the lakes and mountains of Suwa. Epson became prominent in the global environmental movement during the time of its first corporate

* The authors are grateful for interviews and supplementary assistance by the following Seiko Epson engineers and managers: Toshihiko Araki, Alastair Bourne, Takayoshi Fujimori, Masahiro Furusawa, Shuko Hashimoto, Hirokazu Hashizume, Nobuo Hashizume, Yasuhito Hirashima, Haruyuki Imai, Katsuyuki Kanuma, Yutaka Kawaguchi, Itsuo Kobayashi, Saburo Kusama, Yasuo Mitsugi, Toshiyuki Miyajima, Tadashi Miyasaka, Yoshinori Miyazawa, Tsuneya Nakamura, Isamu Namose, Machito Natori, Yoshiaki Ochi, Yoshihiro Ohno, Fujikazu Sugimoto, Toshikazu Sugiura, Katsuji Tanaka, Miwako Uchida, Mitsuhiko Ueno, Kiyohiro Yajima, Yuji Yamazaki, and Hideaki Yasukawa.

pledge to phase out CFCs from precision manufacturing in order to protect the ozone layer. After other multinational companies embraced that pledge, governments gained the confidence to strengthen and accelerate the international agreement to protect the ozone layer—called the Montreal Protocol.[1] Epson's current commitment to preventing climate change is even more ambitious.

On the Sunday evening before our official interview with the company's engineers for this chapter, we are treated to a traditional Japanese dinner at a fine hotel near the shore of Lake Suwa. Our two powerful Japanese hosts, Hideaki Yasukawa and Yuji Yamazaki, punctuate the elegance of the dinner presentation. Hideaki Yasukawa is the Chairman of Seiko Epson. Yuji Yamazaki is Executive Vice President and Senior Director of Environmental Activities and is the son of the former president of Daiwa Kogyo, one of the companies that became Seiko Epson.

(Clockwise from left): Yuji Yamazaki (Executive Vice President and Senior Director of Environmental Activities, Seiko Epson), Hideaki Yasukawa (Chairman, Seiko Epson), (hostess), Sally Rand (EPA), and authors Stephen Andersen and Durwood Zaelke at dinner

Chairman Hideaki Yasukawa is a poised man in his sixties with steel-gray hair neatly cut and brushed back. He is handsome and serious but ready to laugh and to appreciate a compliment about his company's success. He also is a very successful man, who continues to work for love of his company and its vision. Vice President Yamazaki looks as if he is in his late forties, and he has the build of a

1   See Center for International Environmental Law (CIEL), *The Industry Cooperative for Ozone Layer Protection: A New Spirit at Work* (Washington, DC: CIEL, 1994); Stephen O. Andersen, Clayton Frech, and E. Thomas Morehouse, *Champions of the World: Stratospheric Ozone Protection Awards* (EPA430-R-97-023; Washington, DC: US EPA, 1997); and Stephen O. Andersen and K. Madhava Sarma, *Protecting the Ozone Layer: The United Nations History* (London: Earthscan Publications, 2002).

Hideaki Yasukawa and Yuji Yamazaki
discuss climate leadership

middleweight boxer. He is also a successful man who chooses the duty of guiding this multinational company.

Epson's history with CFCs

At dinner we don't talk directly about our genius book project, but rather about Epson's history of leadership in protecting the ozone layer, which began on December 23, 1988. At a time when the original Montreal Protocol only envisioned a 50% reduction in CFCs, Epson pledged to eliminate CFCs within five years, a commitment that many experts considered impossible or even outrageous. The company later accelerated its goal to four years, and still arrived ahead of schedule—completing its Japanese phase-out of CFCs in October 1992, more than three years before the time called for by the final Montreal Protocol phase-out schedule. Epson then completed its global phase-out in May 1993.

In 1970, as head of production engineering, Hideaki Yasukawa had introduced CFCs to Seiko Epson's manufacturing process. "It was my decision to adopt the use of CFCs, so it is ironic that it was my obligation to eliminate them," he says. Yasukawa had decided to replace the chlorinated solvents, such as 1.1.1-trichloroethylene and tetrachloroethylene, and alcohol-based solvents then being used, with CFCs because CFCs are non-flammable, non-toxic, and, according to manufacturer claims at that time, safer for workers, the community, and the environment.[2]

---

2 A translated account of this history is found in Hideaki Yasukawa, *Using Our ODS Elimination Program Experience for the Future* (Seiko Epson Corporation paper; Seiko Epson, March 1995). See also Hideaki Yasukawa, "Eliminating ODSs: Lessons for the Future from Seiko Epson", in Stephen O. Andersen and K. Madhava Sarma, *Protecting the Ozone Layer: The United Nations History* (London: Earthscan Publications, 2002): 203.

During dinner, the Chairman explained his earlier decision. "Using CFCs was the right thing to do at the time because no one had predicted ozone depletion, and I felt that safety should come before cost." During the next decade, CFCs became an integral part of Epson's manufacturing. Once the use of CFC solvents was implemented, every material and every process had to be CFC-compatible. As a result, the manufacturing methods of that time became incompatible with the methods that had been used previously. By 1988, all of the Epson engineers believed that CFCs were indispensable.

> " It is the responsibility of those of us living in the present to pass on to future generations a healthy environment. We intend to do what little we can to continue this effort into the future. "
>
> *Tsuneya Nakamura, President, Seiko Epson Corporation, 1987–91*
>
> quoted in: S.O. Andersen and K.M. Sarma, *Protecting the Ozone Layer: The United Nations History* (London: Earthscan Publications, 2002)

On December 23, 1988, Tsuneya Nakamura, who was President of Seiko Epson at that time, decided to reorganize the company in order to phase out CFCs.[3] His announcement of the establishment of a CFC Phaseout Center received a lot of attention from the Japanese media. One of his goals was to protect the ozone layer and eliminate dependence on unsustainable chemicals. Nakamura also sought to create an organization that could comply with the strict national and global legal requirements that he anticipated would emerge from the Montreal Protocol, which was signed in September 1987.

"I was honestly puzzled by why this decision was seen as unique and why it raised such interest from others," he says. "The reason why I so easily made my decision to phase out CFCs is deeply related to the condition of the environment and to Epson's traditional business policy." At the time when Nakamura made his announcement, none of the Epson engineers had any idea how they could stop using CFCs completely. But Nakamura explained, "It was the only thing to do."

Engineers like a challenge, and once they got over the initial shock of the decision they formed an independent team to tackle the problem. Chairman Yasukawa also explained that Epson believes in setting difficult goals, which result in gaining more wisdom over time. "We expect everyone to scramble. It's a 'start together, achieve together' philosophy that encourages teamwork and stretching of the imagination."[4]

President Nakamura gave the engineers an added incentive when he declared that operations divisions that failed to meet the goal of phasing out CFCs would be closed. This declaration made it clear to every engineer just how seriously the company was taking the problem, and that their work, the stratospheric ozone

---

3   The first report was on December 23, 1988, in *Shinano Mainichi Shinbun* newspaper and one week later it was reported in the national press. On April 7, 1989, Uchihashi Katsuto, NHK Television, asked President Tsuneya Nakamura how CFC solvent could be eliminated; Nakamura replied that Seiko Epson did not know how—he just knew it had to be done.

4   A translated account of this history is found in Tsuneya Nakamura, *The CFC-Elimination Program: Putting our Management Philosophy into Practice* (Seiko Epson Corporation paper; Seiko Epson, March 1995). See also Tsuneya Nakamura, "Eliminating CFCs: Management Philosophy into Practice", in Stephen O. Andersen and K. Madhava Sarma, *Protecting the Ozone Layer: The United Nations History* (London: Earthscan Publications, 2002): 204.

layer, and future generations all depended upon meeting the daunting goal. The stakes were huge!

In the weeks after Epson's announcement, the Japanese and global press printed quotes from suppliers of CFCs, and from manufacturing competitors, who claimed that total phase-out was impossible. This news and controversy attracted a visit from the Montreal Protocol Solvent Technical Options Committee, and Epson used the occasion to issue a dramatic poster proclaiming:

> If only you could see what I have seen, you too would surely act now.
>
> Help pitch in!
>
> Epson to ban CFCs by '93

Posters that Epson printed later were stronger still (Figure 7.1), both in their environmental-responsibility message and in their company resolve:

> The sky's the limit!
>
> The cost of ozone-depleting chemicals is stratospheric.
>
> Seiko Epson is following its successful CFC phase-out campaign with a program to eliminate ozone-depleting trichloroethane by 1993.

and

> Go with the flow!
>
> The current of the times is flowing toward environmentally responsible activities.
>
> We are working together with our global partners and suppliers to accelerate the elimination of ozone-depleting substances.
>
> United, we can beat the ODS.

Although Japan's formidable Ministry of International Trade and Industry initially tried to stifle the company's boldness, Epson's initiative gave courage to the Director of the Japanese Ozone Layer Protection Office which enabled him to pressure the rest of the world into a phase-out of CFCs. At a policy meeting in Tokyo, where industry groups were claiming that CFC solvents were essential, UNEP Executive Director Mostafa Tolba held up the Epson poster and asked rhetorically: "If Epson can phase out, why can't other responsible companies do the same?"[5]

Our dinner closed with Chairman Hideaki Yasukawa explaining how leadership by Epson helped assure the success of ozone protection and the Montreal Protocol, and also laid the foundation for the company's current efforts to protect the global climate:

> The experience gave Epson immense courage to face the challenge of becoming an environmentally clean company. We decided to reduce our own greenhouse gas emissions and to design products to use less energy. A single company does not have the strength to solve this problem alone,

5   The UNEP TEAP's Solvents, Coatings, and Adhesives Technical Options Committee visited Seiko Epson on May 26, 1989, and a week later Dr. Tolba displayed the Epson poster in Tokyo.

Figure 7.1 "The sky's the limit" poster

but nevertheless Epson feels the enormous weight of its responsibility and role as a veteran environmental leader.[6]

The following morning a team of ten top engineers discussed the technological innovations that Epson was pursuing to prevent global warming and to meet its ambitious goal. The Epson goal is to reduce its total corporate energy use by 60% compared to its energy use in fiscal 1997, and to accomplish this by fiscal year 2010. Moreover, Epson plans to do this even if production grows by as much as 10% per year. "Although we all know it is difficult to reduce energy use by 60%, we have no intention of relaxing this target," says Chairman Hideaki Yasukawa.

Nobuo Hashizume, Director of the Global Environment and Safety Policy Office, started the morning session by explaining that success in phasing out ozone-depleting solvents was the beginning of Epson's international cooperation for environmental protection. He says:

6 For the history of how Epson blended engineering and communication, see: Seiko Epson, *Putting the Freeze on CFCs* (Seiko Epson Corporation Public Affairs Department, March 1993).

Many technical experts from other companies and government agencies helped us identify successful technologies for our phase-out. We came to realize a social obligation to help others protect the ozone layer. So we sent our best experts to make presentations at conferences, published technical manuals, opened our factories to inspection by competitors, and even donated patented cleaning technology to the public domain so that costs of ozone protection could be as low as possible.[7]

It is necessary to safeguard environmental conservation technologies with patents, but it is also part of our corporate culture to utilize them as public assets. Our aim is to share our environmental know-how and green technology in combination with our primary role of providing environmental products.

The company is able to live up to these statements because of the impressive work of its engineers, several of whom were present at the session. Only two of the engineers were not wearing ties; one was wearing a conservative sweater, and another had long hair and a coat with a high collar. Ranging in age from their late twenties to late fifties, all exuded a quiet confidence that suggested that they would be comfortable in any of the top companies or universities in Japan, Europe, North America, or anywhere else in the world.

The kinetic power supply

Hashizume and his deputy, Yoshihiro Ohno, were among the young engineers who cut their teeth while meeting the CFC challenge. As part of this challenge, the engineers developed a long-term dream of designing Epson technology powered only by human motion (Figure 7.2). You can feel Hashizume's friendly confidence as he explains that "achieving the new dream of 'power-supply-free' technology will require success on two paths of innovation. One is the path to perfect the technology for generating kinetic power; and the other is the path to drive down the amount of power required to run the computers, printers, cell phones, and other electronic devices."

At seminars organized for our writing team, we hear presentations from engineers working on both challenges. One of these seminars begins with a little background. Epson has been at the cutting edge of the zero-energy electronics revolution. They invented the quartz watch, the Kinetic® watch, and liquid-crystal displays (LCDs) that use very little energy when the display is not changing. The quartz watch exploited the reliability of mineral harmonics to improve timing accuracy. The Kinetic® watch uses the motion of a wrist to generate electricity to power the movement of the watch hands, with power stored in capacitors rather than batteries. Epson LCDs for cell phones show unchanged screen information at

---

7 Epson know-how and inventions to protect the ozone layer were published and distributed worldwide in *CFC-113, 1,1,1-Trichloroethane Alternative Cleaning Technology* (Epson, November 1993).

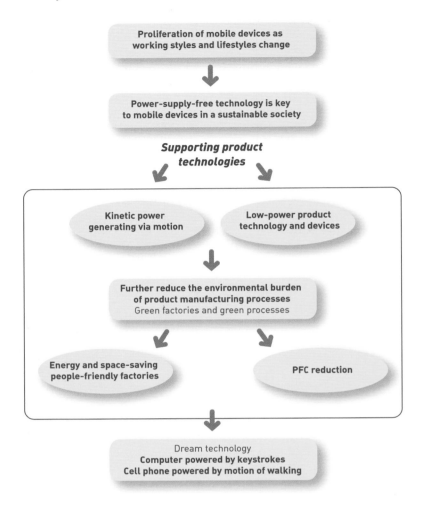

Figure 7.2 Epson's dream of computers powered by keystrokes

near-zero power consumption, so that power is needed only to change the display pixels.

Yasuhito Hirashima, Manager of the Global Environmental Policy Department, says that Epson's innovations in kinetic power supply started in its Watch Operations Division. Hirashima is a man of perhaps 35 years, who wears a flannel shirt buttoned at the top, under a cardigan sweater, with khaki pants. Hirashima is enthusiastic as he describes Epson's work with the Kinetic® watch. The company's latest innovation involves a kinetic system that stores energy in an ultra-capacitor that functions as a secondary battery. This drives the movement of the hands, and never needs to be replaced during the life of the watch. The watch has another innovation—a chip that saves energy by automatically stopping the hand movement ("suspending animation") after 72 hours of inactivity, yet still keeping perfect time. With the benefit of this chip, after a few shakes of the wrist even Rip van Winkle would know how long he'd slept—even after a four-year nap.

A key environmental advantage of the Kinetic® watch is that it requires less material for the manufacture of its energy-storage device, so there is less material to recycle or dispose. You can sense that these engineers have acquired the obligation to "do the right thing" from their highly motivated chairman and president.

Modern society is using an ever-increasing number of mobile electronic devices. From laptop computers to palm pilots and pagers, these devices are adding flexibility to our working habits and enabling us to be more productive. They are changing our lifestyle. Is it possible to power these devices using kinetic energy from our keystrokes and our other motions? Epson believes that, with enough genius, they can accomplish radical improvements in technology for generating power from motion, and that they can reduce power consumption down to its ultimate limits.

The dream is to use Epson kinetic energy-capture technology to generate electricity from the movement of keystrokes and from the motion of carrying portable computers. Solid-state computer components will replace motor-driven disk drives. Energy-consuming fan motors and bulky heat exchangers will be eliminated. Simplified operating systems will function faster and with less power consumption. "No one knows when technology will be ready to create keyboard-powered computers. But our engineers know what our goal is. We will work incrementally until the final breakthrough allows us to market this Earth-friendly product to customers who will appreciate the convenience of always having power," says Hirashima.

Given Epson's commitment to environmental innovation, its demonstrated technical prowess, and the experience and confidence of its engineering team, it is a good bet that they will meet this ambitious goal. Along the road to this ultimate success, they are achieving significant environmental benefits in their own manufacturing operations and in their products. Chairman Hideaki Yasukawa has imposed precise, company-wide goals of reducing total energy use, even as the company continues to grow significantly.

> If all laptop computers sold in 2001 were powered entirely by their keystrokes, 300 million kWh would be conserved—the equivalent of what 40,000 Japanese citizens use in one year.

Current initiatives at Epson

We take lunch at a hillside restaurant overlooking the Seiko Epson Head Office, with Lake Suwa beyond. We see 40-foot-tall (12 m) tourist boats, cast in fiberglass to look like swans (Figure 7.3), creating the illusion that the lake is very close and the buildings on the shore are very small. Later that night we will walk along the shore and marvel at the patience of a half-dozen professional and amateur photog-

Figure 7.3 Large swan boats on Lake Suwa

raphers who are waiting there in anticipation of flocks of real swans. The swans are expected to finish their annual migration right at sunset, creating a masterpiece that combines wings, mountains, light, and water. We suppose that the plastic boats are like decoys luring the swans to the photographers.

Tsuneya Nakamura, who was president of the company from 1987 to 1991 and who is currently an executive advisor, is hosting the lunch and has promised to explain the origin of Epson's commitment to the environment. He offers no talk of the profits of pollution prevention or of responding to the demands of green consumers. Looking back at the no-CFC declaration of 1988, Nakamura explains that, once the damage that CFCs caused to the global environment was known, there was no way the company could continue using them. He believed that one day CFCs would become difficult to obtain, and that it was necessary to establish new technology before this situation occurred.

He goes on to explain that on one hand he was nervous about whether he would be able to convince his employees to cooperate with his plans at the time, when there were no tangible alternatives to CFCs, but how, on the other hand, he was secretly sure that they would cooperate. He believed that this would be a true test of the trust that existed between top management and employees. "I was delighted by the way that our employees responded to my expectations. I am sure that our success in completely eliminating CFCs gave

EPSON MANAGEMENT PHILOSOPHY

❝ Epson is a progressive company, trusted throughout the world because of our commitment to customer satisfaction, environmental conservation, individuality, and teamwork. We are confident of our collective skills and meet challenges with innovative and creative solutions. ❞

EPSON ENVIRONMENTAL PHILOSOPHY

❝ Epson will integrate environmental considerations into its corporate activities and actively strive to meet high conservation standards in fulfilling its responsibilities as a good corporate citizen. ❞

Source: Seiko Epson Corporation, *To the Earth: Green Report Summary*, March 2001

all of our employees a lot of confidence and motivated them to reach other big environmental targets," he says.

He also stresses that being an outsider—a "country-boy company" far from Tokyo—helps protect the company's independence and its ability to set its own goals and go its own way. He sees the dangers faced by the companies in Tokyo and other capital cities when they sometimes delay making decisions by spending too much time debating with each other and governmental and political authorities. 'They may lose track of the main purpose of protecting the environment," he says.

After lunch we turn our attention to energy-saving products. One of Epson's goals is to make products smaller and lighter, while achieving even better performance and adding new features. The company's "energy-saving" technology for next-generation portable equipment embodies that desire. Epson's "power-saving" and "space-saving" designs—with low-voltage, high-density electronics components mounted very close together—enable Epson to create continuously smaller and lighter products.

Haruyuki Imai, who looks like a youthful graduate student, is the next engineer to tell the story of his department's products—the famous Epson printers. But before he starts his technical presentation, he takes time to tell us about the legend of Lake Suwa, which is commemorated by shrines dedicated to a goddess on one shore and to a god several miles away on the other shore. The legend is that they longed for each other, but the waters kept them apart—until winter temperatures froze the lake and created a path. Imai explains that winter may be uncomfortable, but that every natural event exists for a reason, even as an agent of romance. He goes on to tell us that Lake Suwa did not freeze this winter and that he fears this is from global warming.

His team developed the world's most efficient precision printer, measured by the average 24-hour use cycle. Modern printers offering very high resolution must apply ink to an accuracy of within $1/1,000 - 1/2,000$ of an inch. Epson minimizes the energy used by a printer when it is operating by reducing the weight and friction of moving parts, and through software that determines the shortest path to apply the ink. Ironically, however, most of the electricity used by a printer is consumed when it is turned on but is not printing. As a result, Epson engineers designed the printer to use the least amount of energy per day, mindful of the strong demands of customers who do not want to wait for printers to warm up.

When a printer is turned on it goes through a "cold start-up" cycle of motions to clean the nozzles, to align the print heads, and to check for ink in the cartridges. When the printer is idle, there are two power-conservation modes. The "standby" mode occurs just after printing, and has lower power consumption but with the ability to print again instantly. After a short interval, the printer moves into "sleep mode," where the flow of electricity to the motor is all but switched off. The small amount of power used in sleep mode maintains the precise alignment of the print head and paper feed motors, allowing printing to resume relatively quickly without the electricity use and time necessary during cold start-up.

Next, Katsuji Tanaka describes the company's initiative for "compact manufacturing." Epson developed compact manufacturing because it allows increased output without the need to expand existing buildings, which saves space, energy, and money. Compact manufacturing avoids inherently expensive new construction that ties up capital and is frequently subject to delays and cost overruns. When land is not available for company expansion near existing facilities, companies are often forced to build new "greenfield" manufacturing facilities at distant locations that sometimes lack energy-efficient infrastructure such as public transit. When that expansion of manufacturing occurs in urban areas, it has to compete with other demands for housing, transportation, and parks. Tanaka says:

> Before our vision of compact manufacturing, no one cared how large manufacturing equipment was—or how much energy it used. But, once you know how energy is being used, your eyes are opened to new possibilities. New Epson compact manufacturing machines and facilities are smaller, more energy-efficient, more reliable, and even have faster production rates. Today we measure engineering excellence by the amount of energy we save.

Epson's compact manufacturing is not simply concerned with raising productivity. The company is also aiming to improve operability and reduce the burdens on the operators. One successful example of this is how it reduced the distance that operators had to walk by more than half. When considering operability, Tanaka's team found that it is more efficient to work for short periods standing up. However, when a job lasts for more than half an hour, sitting down reduces the strain on the body and is more efficient. Designed for improved operability, many Epson manufacturing processes can be performed either standing up or sitting down, depending on whether operators need to remain in the same place for an extended period of time.

> ❝ Tending to world environmental problems such as protecting the ozone layer, preventing global warming, and developing counter-measures for acid rain is the common responsibility of all of us on Earth. It is a task that should be confronted by concentrating the wisdom of mankind. These environmental problems now transcend national borders and countries; businesses and even individuals must behave as though each problem is their own. ❞
>
> *Tsuneya Nakamura, President, Seiko Epson Corporation, 1987–91*
>
> quoted in: S.O. Andersen and K.M. Sarma, *Protecting the Ozone Layer: The United Nations History* (London: Earthscan Publications, 2002)

In addition, Epson's miniature and desktop manufacturing machines, which are up to ten times smaller than their earlier counterparts, are being integrated into existing and new assembly lines. Epson's compact manufacturing places new, highly reliable and moveable manufacturing equipment together more closely. Production capacity has been doubled on the existing floor space at prototype facilities. These new machines are so reliable that they no longer need adjacent open space for disassembly and repair. Compact manufacturing has also already reduced energy use for materials handling.

Another Epson manufacturing process has also saved space and energy, this time in the manufacture of quartz devices. Indeed, developing and introducing space-saving production lines has *halved* both space and energy; and production flexibility has been added. "Compact manufacturing allows us to respond agilely to sharp increases in demand without the need to build new facilities or expand existing ones. Epson earns a lot of money from such inventiveness and manufacturing agility," says Yoshinori Miyazawa, Head of the Environmental Product Committee.

The company's Semiconductor Operations and Display Operations Divisions, which account for 65% of energy consumption and emissions, are not only developing compact manufacturing but are also reducing the number of production stages in manufacturing. They will have created entirely new processes by 2010.

Toshikazu Sugiura, who is in his late twenties or early thirties, is one of the new generation of Epson engineers already making extraordinary contributions to semiconductor manufacturing. He is wearing a white shirt with a green lab coat. His hair is parted in the middle and hangs down over his forehead to his eyes. He is an inventor in the Semiconductor Operations Division, where he and his colleagues are struggling to complement Epson's energy-efficiency target with an aggressive goal to reduce PFC emissions from their semiconductor production process. PFCs are the most potent and persistent greenhouse gases, but are currently essential for the manufacture of semiconductor and LCD devices. Without reductions in PFC emissions, however, the climate protection achieved through Epson's energy savings could easily be overshadowed, or even made irrelevant.

Energy-efficiency initiatives by top management have yielded an extraordinary pay-off at the Display Operations Division, according to Toshihiko Araki, General Manager. He says:

> At first we were all afraid of the ambitious energy-efficiency goals to protect the climate. Our president told us to "think outside the box" and to inspire young engineers. He was right. Our environmental ethic allowed us to imagine the technology used to develop cell phone displays that consume little energy when the display pixels are unchanged. Thanks in large part to this technology, today we are one of the largest suppliers of cell phone LCDs in the world.

Epson has made billions of dollars in sales because the energy efficiency of its cell phone displays translates into longer battery life.

Taken together, the Epson product and manufacturing energy-efficiency goals—and the PFC emission reduction goals—are among the most ambitious in the world. Epson will match its energy efficiency goals by fiscal 2010 by also reducing emissions of PFCs from semiconductor and display manufacture by 60% below fiscal 1997 emissions. These absolute reductions in emissions are all the more extraordinary because Epson expects its sales growth to continue to increase. The company's average annual increase between 1997 and 2001 was 10%.

Without the company's global warming strategy, PFC emissions from rapidly expanding semiconductor and display production would have undergone a projected tenfold increase by 2010 compared to the 1997 base year. That means that a 60% absolute reduction is actually a 97% reduction per unit output! Epson is one of the few semiconductor companies in the world with a goal of basically zero PFC emissions. This is reminiscent of their stance on CFCs a decade ago. Imagine the reduction of traffic on our roads—and the reduction in pollution and climate change—if automobile manufacturers, city planners, and customers could work together to match this performance. And with an atmospheric lifetime of 5,000–10,000 years, these reductions in PFC emissions may benefit society even more.

Epson time-line

1942 ● Daiwa Industry (Seiko Epson's predecessor) founded in Suwa.
　　　● Daini Seikosha relocates part of its operations to Suwa from Tokyo.
1963 ● First portable quartz chronometer.
1964 ● Seiko Group appointed the Official Timer of Tokyo Olympics.
1968 ● Digital electronic printer (EP-101) introduced.
1969 ● First quartz watch.
1973 ● First six-digit LCD digital quartz watch.
1975 ● First multifunctional digital quartz watch.
1978 ● First twin quartz watch with accuracy within less than 5 seconds per year.
1982 ● Shinshu Seiki (Seiko Epson's subsidiary company; founded in 1961) name
　　　　changed to Epson Corporation.
　　　● First TV watch.
　　　● First hand-held computer.
1983 ● First pocket color LCD TV.
1988 ● Epson is first global manufacturer of precision products to pledge CFC
　　　　phase-out (December).
　　　● First "Environmental Benchmark Year" for tracking environmental
　　　　performance.
　　　● First Kinetic® quartz watch.
　　　● First intelligent analog quartz watch with central processing unit (CPU).
1989 ● First liquid-crystal video projector introduced.
1990 ● First pager watch (Seiko Receptor).
1991 ● The CFC Phaseout Center transformed into the Environmental Affairs
　　　　Office with responsibility to lighten the burden on the environment by all
　　　　Seiko Epson business operations.
1992 ● Epson is first global precision product company to complete domestic CFC
　　　　phase-out (October).
　　　● Epson is first global company to pledge phase-out of 1,1,1-trichloroethane
　　　　(phase-out to be completed 1993).
　　　● Epson wins EPA Stratospheric Ozone Protection Award.
1993 ● Epson CFC phase-out at all global operations (May); phase-out in major
　　　　suppliers (July).
　　　● Epson is first global company to phase out 1,1,1-trichloroethane worldwide
　　　　(November).
　　　● Tsuneya Nakamura and Yasuo Mitsugi win EPA Stratospheric Ozone
　　　　Protection Award.
1994 ● Hideaki Yasukawa wins EPA Stratospheric Ozone Protection Award.
1995 ● Epson Hong Kong wins EPA Corporate Stratospheric Ozone Protection
　　　　Award.
　　　● Yuji Yamazaki and Kaichi Hasegawa win EPA Stratospheric Ozone
　　　　Protection Award.
　　　● Epson Portland wins EPA Corporate Stratospheric Ozone Protection
　　　　Award.

1997
- EPA Best-of-the-Best Corporate Stratospheric Ozone Protection Award to Seiko Epson.
- EPA Best-of-the-Best Individual Stratospheric Ozone Protection Award to Tsuneya Nakamura, Hideaki Yasukawa, and Kaichi Hasegawa.

1998
- "Second Environmental Benchmark Year" for strengthening overall environmental measures (ten years since the no-CFC declaration).

1999
- First suspended animation watch (Kinetic auto relay).
- First publicly available environmental report disclosing environmental performance.

2000
- Testing and introduction of Epson Ecology Label (conforms to Energy Star, but is more stringent in energy conservation when in power-off, and meets other environmental standards for recyclable design).

2002
- Eliminate almost all of the lead in the solder used on circuit boards, with some exceptions (March).

2004
- Goal of 100% green purchasing for production materials.

2011
- Goal of 60% reduction in energy consumption and emissions (1997 base year).
- Goal of 60% reduction in high-GWP greenhouse gas emissions (PFCs and SF$_6$, 1997 base year).

8

ST Microelectronics
Turn on, plug off *

Using the power of semiconductors to eliminate energy waste in battery chargers and other consumer products, ST is cutting electricity consumption dramatically.

ST President and CEO,
Pasquale Pistorio

ST Microelectronics, the third largest semiconductor manufacturer in the world, designs, develops, manufactures, and markets one of the largest portfolios of products, systems, and applications in the industry. The company has 18 primary manufacturing sites around the world and over 40,000 employees. Operating in a world of extreme competition, hyper-innovation, and boom-and-bust business cycles, ST Microelectronics stands out for its attention to the environment and its corporate success.

"Environmental commitment is not an option, it's a moral and social obligation, and as a company with 2001 global sales of over $6 billion, we are committed to be a leader in this field," says Pasquale Pistorio, ST President and CEO.[1] ST is not only implementing this philosophy in its manufacturing processes and products, but it has simultaneously risen over the past

* The authors are grateful for interviews and supplementary assistance by ST engineers and managers Claudio Adragna, Ivonne Bertoncini, Giuliano Boccaletti, Fabio Borri, Carlo Cazzaniga, Vincenzo Daniele, Eugenio Ferro, Enrico Galbiati, Francesca Illuzzi, and Roberto Zafalon.
1 "Message from the President", *ST Microelectronics Corporate Environmental Report 2000*.

decade to an elite technical and environmental status among global semiconductor manufacturers.

Semiconductor integrated circuits and discrete devices act as the brains for advanced electronic controls and devices, and are ubiquitous in consumer and industrial applications. Personal computers, for example, contain several different types of semiconductor devices, and these devices are also found in vehicles, appliances, telecommunications equipment, and building and industrial processes. The services and applications enabled by semiconductors enhance and enrich all of our lives. There are many types of semiconductors, with varying complexity and design features. Depending on the type and performance features, semiconductors

Figure 8.1 Computers save energy in all products, including cars

can have enormous influence on the functionality, reliability, access time, power consumption, and cost of products.

Chapter 4 on Energy Star explains why energy efficiency is often neglected in the design of electric appliances, and how industry–government partnerships are aggressively labeling and promoting energy-efficient products to allow informed customers to choose models that save money and protect the environment. Although the EPA is constantly adding new products to the Energy Star program, many products are not yet labeled. New technology from ST provides another opportunity to significantly reduce the cost of appliance ownership and protect the atmosphere and its climate.

ST pledges to be "climate-neutral" by 2010

ST Microelectronics is one of only a dozen or fewer companies in the world pledged to be "climate-neutral" by 2010.[2] Climate-neutral companies:

- Reduce the energy consumption of their manufacturing processes, buildings, and transport to minimal levels

- Procure as much of their energy as possible from green sources that contribute minimal greenhouse gases (biomass, hydroelectric, solar, and wind)

- "Offset" the net quantity of greenhouse gas emissions with investment in energy conservation, renewable and alternative energy sources, and carbon sequestration (tree planting)

Thus, climate-neutral companies strive to achieve net zero impact on the climate. Through climate-neutrality, ST expects to save $ 900 million in the period between 1994 and 2010. For ST Microelectronics, however, neutrality is not enough—the real test is in the environmental consequences of the products it invents and makes. The company's semiconductor products must go beyond climate-neutrality and serve as a cornerstone for efforts by its customers to improve energy efficiency. This combination of environmental responsibility and efficiency empowerment is captured in Pistorio's motto "do no harm, do much good."

The message of environmental protection is embedded in the daily lives of ST employees—from how they get to work to what is available to eat in the cafeteria, and in how their job performance will be evaluated. The goal of protecting the environment carries significant weight in the design, production, and disposal phases of ST products. And, potentially most important of all, ST puts a great deal of consideration into how environmental performance, in particular the use of energy, of ST product applications can be enhanced.

---

2   Companies striving to be climate-neutral can get more information from the Climate Neutral Network at www.climateneutral.com.

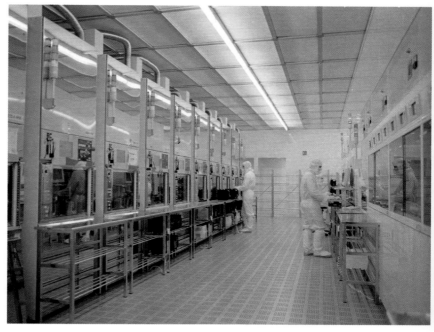

Figure 8.2 Clean room where semiconductors are manufactured

Evolutionary energy savings from semiconductor innovation

Semiconductor manufacturers have greatly decreased energy consumption per unit of calculation.[3] Energy has been saved by increasing the efficiency of components such as computer monitors, by replacing moving components such as hard drives with solid-state memory, and by creating software that reduces energy

3   The standard unit of measurement is "energy dissipation" per "elemental operation," which is defined as the operation performed by a single device within one clock period. Because the elemental area on the chip is represented by the square of the process critical dimension (CD), the complexity is proportional to the ratio of the circuit area divided by this elemental area: area/(CD)^2. For digital devices, the formulation of the index becomes: power dissipated by the circuit × (CD)^2/area × clock frequency, which is proportional, but not strictly equal, to the energy spent per device and per clock cycle. The formulation for stand-alone memories, where the complexity is independent from the size of the memory, is simple: power dissipated by the circuit × clock frequency/number of bits per word.

consumption when full power is not needed. For example, ST reduced energy consumption per calculation in its erasable programmable read-only memory (EPROM) semiconductor devices by a factor of ten over a period of five years. EPROMs are used in a vast array of home and office electronic equipment, including printers, fax machines, and stereos. This efficiency is incorporated automatically into all products that use this device. Any increase in device energy efficiency also reduces the amount of heat that must be dissipated.

Mobile phones, laptop computers, digital cameras, MP3 players, and an ever-increasing range of portable equipment make life considerably more convenient and more fun. However, the proliferation of portable products has meant that battery charging and alternating current (AC) adapters, which once had an insignificant environmental impact, now have a significant one. AC adapters and battery chargers in use today consume several TWh of electrical energy per year, and their number will continue to increase.

ST semiconductor technology simultaneously achieves the product performance challenge of making batteries last longer to better satisfy customers, and the environmental performance challenge of making chargers more energy-efficient.

Battery-powered equipment in use today also consumes large amounts of power when it rests in "standby" mode. Electronic products that consume power when they are not in use are sometimes called "electricity vampires," because they suck energy from the planet. Technology to avoid this waste was not available previously because engineers have not been challenged to design or build power-saving features, and because product manufacturers believed that even a very small increase in cost could not be recovered through a higher selling price. Consumers never really know how much energy a product will use and often do not appreciate that electricity may be consumed even when the appliance is turned off.

Rechargeable batteries are important to the commercial success of all portable electric devices, including electric and hybrid cars. Battery designers strive for (1) high capacity and reliable voltage at likely temperatures and level-of-charge, (2) long life, and (3) materials that are safe to dispose or recycle. The useful life of most batteries is longest if consumers use the device until the battery is fully discharged and then recharge it completely without overcharging it. However, most customers frequently "top up" the charge before the batteries have fully drained to guard against the running-out of power, or schedule recharging over night when they are less likely to need the device. Frequent topping-up of batteries without allowing them to fully drain can create a condition called "battery memory," where the battery loses its capacity to hold a full charge and drains quickly when used.

As long as they are plugged into the electric outlet, today's cell phone battery chargers consume electricity whether or not the cell phone is connected. Until now, the only way to avoid this waste, and the pollution it causes at the electric power plant, was to unplug the charger from the wall as soon as charging was completed. Even the most environmentally conscientious consumers are faced with the complication of variable charging times and the inconvenience of periodic checking, particularly when most products are set to charge overnight. Most people just leave them plugged into the wall regardless of whether they are charging.[4]

But when product designers and consumers look at this waste of energy from ST's "every milliwatt counts" perspective, it is easy to recognize that the environmental impact is huge and that technology that pushes toward eliminating this waste offers enormous social benefit.

ST's "near-zero" emissions tolerance policy

To see that cell phone battery chargers consume this unneeded energy, just touch the charger when it's plugged in, and the heat tells you that energy is being consumed. ST-empowered cell phone battery chargers, on the other hand, have "near-zero" consumption of electricity when not in use.

So how has ST made this technological advance? ST's unique battery charger architecture and state-of-the-art components have dramatically improved charging efficiency and power consumption during "no-load" conditions when the charger is plugged in but not charging a battery (see Table 8.1 and Figure 8.3). Many designers of cell phones and other portable electric devices are taking advantage of new battery designs and "intelligent battery chargers" to improve performance and convenience.

ST smart chargers could save 1,560 GWh for the 300 million mobile phones sold worldwide in the year 2000 if phone chargers were left plugged in—enough to power three typical European towns of 100,000 people. If the phones are only plugged in at night, 152 GWh could still be saved. And CO_2 greenhouse gas emissions are reduced by 1,000 tons for every GWh saved.

---

4   ST has also created products and architectures to drastically reduce electricity consumption when the load is removed. Many products with remote controls are always on whenever the appliance is plugged in, constantly consuming electricity in order to be ready to respond to commands from the remote control. ST semiconductor's power-saving design drastically reduces the standby power consumption of products using remote controls. This technology can be scaled to suit many appliances, including air conditioners, fans, TVs, VCRs, DVDs, audio amplifiers, CD players, and video games. ST's goal is zero standby power.

	Charger always plugged in (Watt-hours per week)	Charger plugged in at night (Watt-hours per week)
1996 Typical charger*	114.8**	11.20†
2002 ST smart charger††	14.8‡	1.44‡‡
Weekly savings with ST	100.0	9.76

* Battery capacity: 1 Ah 7.2 V
 Ein = 13 Wh, energy sunk from the electric outlet for charging a battery completely
 tc = 2 h, charge time
 Ploss = 0.7 W, battery charger power losses in no-load condition

** Eint = 2 × Ein = 26 Wh, energy used for two chargings in a week
 tp = 164 h, plug time in no-load condition
 Eloss = Ploss × tp = 114.8 Wh, energy lost in a week

† Eint = 2 x Ein = 26 Wh, energy used for two chargings in a week
 t'p = 16 h, plug time in no-load condition
 E'loss = Ploss × t'p = 11.2 Wh, energy lost in a week

†† Battery capacity: 1 Ah 3.6V
 Ein = 7 Wh, energy sunk from the electric outlet for charging a battery completely
 tc = 2 h, charge time
 Ploss = 0.09 W, battery charger power losses in no load condition

‡ Eint = 2 × Ein = 14 Wh, energy used for two chargings in a week
 tp = 164 h, plug time in no-load condition
 Eloss = Ploss × tp = 14.8 Wh, energy lost in a week

‡‡ Eint = 2 × Ein = 14 Wh, energy used for two chargings in a week
 tp = 16 h, plug time in no-load condition
 Eloss = Ploss × tp = 1.44 Wh, energy lost in a week

Table 8.1 Improved power consumption for charging batteries

Figure 8.3 ST chip and circuit board achieve near-zero power consumption

Economic feasibility of energy savings requiring action by manufacturers

ST estimates that for most products that use semiconductors, its power-saving technology can be applied at little or no cost. Moreover, the application of ST technology will generate power savings that will pay back any small costs that are incurred. The costs break down as follows:

- For 10% of products, power-saving technology already included in ST chips could be empowered at no cost in manufacturing and with minimal design cost.

- For 75% of products, power-saving technology currently available in ST chips could be empowered at small additional cost in manufacturing and design. Added cost would be recovered in electricity savings within one year of purchase.

- For 90% of products, new power-saving technology could be promptly designed and supplied by ST at modest additional cost in manufacturing and design, but recovered in electricity savings within five years from purchase.

Altogether, currently or nearly available chip power-saving technology could reduce power consumption by amounts between 80% and 95%. If implemented worldwide, this would reduce electricity consumption by between 50% and 90%. Table 8.2 outlines how this is accomplished. Circuits are simplified, made more energy-efficient, and are put in power-saver mode when not in use.

Moral obligation leads to competitive advantage

ST Microelectronics came about in 1987 as the result of the integration of the French manufacturer, Thomson Semiconductors, with SGS Group, the only Italian microelectronics company at that time. Even before climate protection was a major global concern, under Pistorio, ST Microelectronics developed its socially motivated philosophy. It was based on the strong conviction that environmental protection achieves ST's triple bottom line of cost saving, reputation, and sales growth, as well as the benefit of attracting and retaining top talent. The basic foundation of the ST environmental philosophy is grounded in the 1987 Brundtland Commission Report: corporate sustainable development must preserve the future by not destroying the present for economic development.[5] ST also links environ-

---

5 In 1987, the Brundtland Report of the UN World Commission on Environment and Development, also known as *Our Common Future* (New York: Oxford University Press, 1987) alerted the world to the urgency of making progress toward economic develop-

Level	Application power saving (%)	Chip aspect to be optimized for life-cycle energy efficiency (device and all peripherals)
Physical*	5–10	Process technology, Multi Vth (threshold voltage), SoC (system on chip)
Device*	10–20	Circuit/layout
Gate/logic*	20–30	Circuit mapping Low-power low-voltage cell library
Internet Protocol (IP)/block*	30–50	As-needed clock-gating Voltage scaling
Architectural†	40–80	Partitioning, allocation Power sharing
System†	80–95	Algorithms, avoid waste Software compilation, operating system

* Available to device customers now

† Technically feasible, and available to device customers if requested

Table 8.2 Chip power-saving technology

mental protection closely with total quality management principles. It links them so closely in fact that Corporate Vice President Georges Auguste is the Director of Total Quality *and* Environmental Management.

There is plenty of evidence large and small of the sincerity of ST Microelectronics' environmental ethic: the first day of each quarterly three-day executive meeting is devoted to environment and total quality management; 2% of capital investment is for green measures. Each year, all managers receive copies of the environmental scorecard *The State of the World* (published by the Worldwatch Institute), and are expected to reflect on how the company can contribute to sustainable development. *The State of the World* is our bible," says Vice President Auguste. "Managers study it and can cite chapter and verse."

ST is a company that has a long-term vision with attention to environmental detail. For

> " Ethical values, responsibility, and ideals are an important base for motivating people to enhance their capacities as individuals and as members of our organization. "
>
> *Pasquale Pistorio, President and CEO of ST Microelectronics*
>
> "Message from the President", *ST Microelectronics Corporate Environmental Report 2000*

ment that could be sustained without depleting natural resources or harming the environment. Written by an international group of politicians, civil servants, and experts on the environment and development, the report provided a key statement on sustainable development, defining it as: "development that meets the needs of the present without compromising the ability of future generations to meet their own needs."

example, employees receiving company cars must choose from a list of the most "green" cars available locally. Cash refunds are given if you choose one of the most environmentally sound cars. President and CEO Pistorio chose the gasoline–electric hybrid Toyota Prius. You can see other hybrids driving around ST production sites making deliveries. (See Chapter 6 on Honda for more discussion of hybrid cars.) Another small, but meaningful, action at the Agrate Brianza Italy research and production facility allows employees to learn about sustainable agriculture from the head of the cafeteria; and they can choose organically grown items off the daily menu.

The ten-point "decalogue" of ST's environmental goals has been available to the public since 1995. In addition to goals for reducing, re-using, and recycling, it lays down time-lines and quantitative milestones. The decalogue is communicated to all employees and is distributed widely to external stakeholders including governments. The decalogue is another brainchild of ST President Pasquale Pistorio. Mr. Pistorio credits his environmental fervor to his eldest son Carmelo. "I am proud that Carmelo is a philosopher and liberal environmentalist and he is proud that I am an industrialist and liberal environmentalist—even though we have not always agreed," says Pistorio. "He persuaded me that the environment should be a priority, and I proved to him that environmental ethics can guide a company to financial success."

"Every thoughtful executive knows that the industrialized world is destroying the planet. What distinguishes ST is that we care enough to do something about it," Pistorio continues.

> ST's culture is nurtured by two strong beliefs: first, ecology is free, meaning prevention is cheaper than correction. Second, companies that anticipate future legislation rather than react to it are more efficient. Since eco-efficient corporations use, and pay for, fewer natural resources, they are intrinsically more profitable than others. Shareholders' value is not threatened by corporate, social, and environmental responsibility; on the contrary, being good citizens amplifies stakeholder value and return to investors.[6]

Being "green" makes good business sense

Despite his strong environmental vision, Mr. Pistorio never forgets he is a businessman. He implements his philosophy through a number of simple yet highly effective management directives. First, don't spend unless there's a return. Capital turns over quickly in the semiconductor industry, usually within three years. Semiconductor manufacturing is also a turbulent industry with dramatic expansion and contraction cycles. For ST, return can be financial gain, such as an energy payback of two to three years, or a strategic return that better positions the company

6 "Message from the President", *ST Microelectronics Corporate Environmental Report 2000*.

for change, improves the global mobility of production processes, or anticipates future legislation.

ST applies the most stringent environmental regulations applicable in any country where it has production to all sites globally. For example, for waste-water discharge, fluoride emission limits range from 15 mg per liter in France to 6 mg per liter in Italy. Thus, ST meets the most stringent 6 mg per liter limit at all locations where it operates. A second directive is to be proud of progress. Environmental accountability includes continuous monitoring of progress, including periodic audits of all sites worldwide. Corporate environmental audits are carried out every 18 months at every site to ensure that environmental procedures are followed, and to monitor performance and progress on achieving goals. Results are shared company-wide to encourage competition and to provide recognition and reward.

The third directive, "be first," is a clear expectation, and it is simple to measure. ST benchmarks environmental progress against itself and against leading companies the world over in order to learn from experience and to demonstrate that its results equal or exceed the best environmental performance of other companies. One "first" mandate was to be the first semiconductor company to achieve the Eco-management and Audit Scheme (EMAS) validation and the ISO 14001 certification in each country of operation.

Environmental managers still debate hotly whether ST or a competitor should be credited with receiving the first ISO 14000 series certification in North America. In February 1996, ST completed ISO 14000 certification requirements using draft standards that ultimately did not change. So, when the final ISO 14000 was issued in September 1996, the third-party certification body converted the previous certificate, keeping the original date.[7] ST therefore correctly cites itself in promotional materials as being the "first." The competitor, who received the first certification *after* the final requirements were issued, also refers to itself as first—to the continued consternation of ST.

The rewards of corporate virtue

Just as the company has continued to realize environmental goals, ST has delivered positive economic results by steadily moving up in the Dataquest™ worldwide ranking of semiconductor companies—from 13th in 1992 to 8th in 1999 to 3rd in 2001.

A final illustration of ST's commitment to the environment—and a conscious part of its strategy for accelerating corporate achievements—is the industry leadership role of work collaboratively with competitors to set and advance industry-wide environmental goals. ST wants other world companies to learn from its environmental experience, and it hopes that others will follow the same direction toward protection of the environment.

7   Joseph Cascio (ed.), *ISO 14000 Handbook* (Milwaukee WI: CEEM Information Services with ASQC Quality Press, 1996).

The semiconductor industry is fiercely competitive, but also somewhat unique in its willingness to cooperate on issues of mutual benefit, such as the environment. This willingness comes from both a genuine commitment to environmental stewardship and a certain amount of enlightened self-interest. The costs and risks of the semiconductor business are enormous; thus it is attractive and smart business to accept the benefits of sharing information and collaborating on non-competitive issues such as the environment.

In the early 1990s the industry faced a new and technically daunting challenge. The manufacture of semiconductors was becoming increasingly dependent on a group of fully fluorinated chemicals called perfluorocarbons (PFCs). These gases are used both for etching silicon wafers and for the plasma-enhanced cleaning of reaction chambers used in chemical vapor deposition.

PFCs have unique physical and chemical properties that make them ideally suited to the needs of semiconductor manufacture—and very difficult to replace. They are also mostly non-toxic and non-flammable, which are characteristics that normally delight environmental, safety, and health managers. These gases, however, are also very potent and persistent greenhouse gases. PFCs have several thousand times more heat-trapping impact in the atmosphere than CO_2, and they have atmospheric lifetimes of hundreds to thousands of years—leading to essentially irreversible atmospheric impacts. Initially the industry only used small amounts of these chemicals, but, with production growth rates well over 12% per year in the early 1990s, it would not take long before their emissions would be significant.

The semiconductor industry's actions over the past decade are easily worth a chapter on industry "sector" genius, but ST's role in motivating the industry to take collective action has been especially important. Dr. Fabio Borri, ST's recently retired Corporate Director of Environment Strategies and International Quality Programs, is remembered as genial but relentless in his pursuit of industry action on PFC reductions. Supported by his managers and ST process engineers, he championed both the research needed to advance PFC reduction efforts and the need for the industry to make a public commitment to reduce emissions. During the critical time when the industry considered PFC issues, Borri served as head of the European Semiconductor Industry Association (ESIA) PFC Task Group and as chairman of ESIA delegation to the World Semiconductor Council (WSC) Environmental Safety and Health (ESH) Task Force. These Task Forces wrestled with the question of whether it was technically feasible to reduce emissions of PFCs, and whether the industry would be willing to set a global emissions reduction goal.

Borri is the personification of the European gentleman, so no one could ever really get mad at him, but many who participated in the WSC meetings readily acknowledge that his persistence was formidable and became an important factor in reaching consensus to take action. In April 1999, WSC members announced a goal of reducing PFC emissions by at least 10% below the 1995 baseline level by year-end 2010. This was the first, and still is the only, industry to agree voluntarily to reduce global climate-change emissions worldwide.

Another reason ST gives for embracing the environment as a moral obligation is that company values must be in line with personal conviction. Borri's interest in the environment is clearly as much personal as it is professional. When the WSC ESH Task Force met in Hawaii, the one "tourist" thing he wanted to do was visit the

Figure 8.4 Semiconductor manufactured by a process emitting fewer
PFCs into the atmosphere

Moana Loa Observatory where data is collected on atmospheric concentrations of greenhouse gases. Not only did he go to the trouble of getting permission through various US government officials to visit the isolated volcano-top station, but he also proudly shows off his trip memorabilia, which he has collected in an album. The album includes neatly labeled photos of the site and its various instruments, plus all of the correspondence and maps that he accumulated while arranging the visit. The site rarely has "general-interest" visitors, and certainly doesn't have a visitor's center. In fact, there wasn't even any staff there on the Saturday of Borri's visit. Nonetheless, he says it was the most fun he'd had in a long time, and a genuine thrill to see a place that is dedicated to advancing the science of climate change.

ST's relentless pursuit of environmental values and practices is paying large dividends for climate protection and its bottom line. By manufacturing products that are less power-hungry, and which optimize energy efficiency during their application, ST matches the full potential of semiconductors to the power of electronic technology. ST technology is opening doors to previously ignored energy-saving opportunities, and it promises even greater advances in the future.

ST Microelectronics time-line

1987   ● ST formed from two companies, SGS Microelectronics and Thomson Semiconductor.

1988   ● First sample of 1 Mbit EPROM.

1993   ● CFCs eliminated from all processes.
● First ten-point "decalogue" quantifies ambitious corporate goals.
● Dedicated Product Group (Agrate, Italy) exceeds $1 billion in sales; 1993 Agrate Brianza wafer fabrication exceeds 1 billion wafers produced.

1996   ● ST Asia Pacific exceeds $1 billion in sales.

1997   ● All 17 manufacturing sites are EMAS-validated and ISO 14001-certified.
● European Quality Award.
● Over 10 million MPEG decoder integrated circuits sold.

1998   ● Leadership in forming World Semiconductor Council PFC Task Force
● French Ministry of the Environment and French Chamber of Commerce prize for Gestion Environnementale.
● Trophées Entreprise Environnemental.
● Award and Special Commendation from the Jury: European Better Environmental Award For Industry—Managing for Sustainable Development.

1999   ● Dow Jones Sustainability Group Index ranks ST among the top 200 of nearly 3,000 companies (one of six "selected industry group leaders").
● Energy, PFC, and chemical roadmaps.
● US EPA Corporate Climate Protection Award "For extraordinary leadership in reducing greenhouse gas emissions from semiconductor manufacturing".

1999 ● Second environmental "decalogue" issued, including goal of making ST a zero CO_2 emission company by the year 2010.
● Singapore Quality Award.
● Malcolm Baldrige Quality Award.

2000 ● Electricity and water consumption reduced by 29% and 45%, respectively, versus the 1994 (equal production rate) baseline.
● Inauguration of new 8 in (200 mm) wafer fab in Rousset, France.

2001 ● Akira Inoue Award to Pasquale Pistorio "For outstanding achievement in environment, health, and safety in the semiconductor industry".
● New state-of-the-art assembly plant opened in Morocco.

2002 ● *Tomorrow* Environmental Leadership Award.
● US EPA Individual Climate Protection Award to Fabio Borri "For passion and persuasion in industry voluntary programs to protect the Earth for future generations".

Trane
It's about making green*

A breakthrough in energy efficiency for building and manufacturing process air-conditioning chillers raises the bar on environmental performance and cost savings.

Can a company be green and make money? Trane's founding Director of Environmental Affairs, Eugene Smithart, likes to say, "The *only* way for a business to be green is to make green."[1] Known as "Smitty," he adds, "At Trane, we firmly believe the future belongs to those who develop technology that is good for business as well as the environment—technology that delivers the highest efficiency and the lowest emissions on a documented, sustainable basis. We will continue to prosper and grow as a business if we help our customers protect the environment."

Trane was founded in La Crosse, Wisconsin, in the late 1800s by a Norwegian plumber, James Trane. He built the business foundation that his son Reuben would eventually take into the heating and cooling business. Trane developed a *unique corporate culture*, thanks to James and Reuben Trane and the Wisconsin work ethic they embodied. James insisted that Trane serve its customers first. His son Reuben insisted that the company be a leader in technology. Their philosophy was to hire engineers for every position in the company, even the sales force.

Today, Trane's executives continue to be engineers selected for their engineering and business prowess. Trane business executives understand the value of product innovation and are not shy about going to the chalkboard during meetings with their engineers to draw their own graphs and formulas to illustrate their ideas. *The*

*   The authors are grateful for interviews and supplementary assistance by Trane engineers and managers Gerald E. Arndt, Lee R. Cline, James A. Dudley, David Eber, Paul R. Glamm, Bryon Hamm, Terry Pickett, John H. Roberts, Todd W. Smith, Eugene "Smitty" Smithart, Brian Sullivan, Mike Thompson, and Jim Wolf.

1   In January 2003, Eugene Smithart moved to Turbocor, where he is Vice President of Sales and Marketing. Turbocor is on our "Honor Roll" of companies that will be profiled in subsequent editions of this book.

s one language: engineering, and engineering means a constant
tter way to build products that best serve their customers and the

Creating the right atmosphere

Until recently, the air-conditioning industry depended on CFC-11 and CFC-12. Before
the invention of CFCs, refrigerants were either flammable or toxic, or both. A leak
of the most common refrigerants—sulfur dioxide and ammonia—typically required
rapid evacuation of homes and buildings, and people who came into contact with
those substances suffered from vomiting, burning eyes, and painful breathing. Fortunately, though, these accidents rarely resulted in death. Leaks of the less common refrigerant methyl chloride, in contrast, frequently did result in deaths.

From their invention by Thomas Midgley at the General Motors Laboratory in 1928, until 1974, CFCs were considered perfect in every known way—because stratospheric ozone depletion was neither understood nor anticipated.[2] CFCs are chemically stable, non-corrosive, non-flammable, odorless, colorless, energy-efficient, inexpensive, and have very low toxicity. Refrigeration and air-conditioning equipment manufacturers and their customers came to think of CFC as a "wonder gas."

Then, in 1974, Mario Molina and F. Sherwood Rowland warned that CFCs would deplete the stratospheric ozone layer that protects life on Earth. In the next two decades the Molina–Rowland hypothesis was scientifically verified; the 1987 Montreal Protocol (that governs the use of CFCs and other ozone-depleting substances) was signed, ratified, and entered

> ### THE OZONE LAYER
>
> The ozone layer is a concentration of ozone molecules in the stratosphere from 9 to 22 miles (14–35 km) above the Earth's surface. Ozone is a molecule containing three oxygen atoms. Normal oxygen, which we breathe, has two oxygen atoms. Ozone is much less common than normal oxygen. Out of each 10 million molecules in air, about 2 million are normal oxygen, but only three are ozone. About 90% of Earth's ozone is located in the stratospheric ozone layer. This ozone is a naturally occurring gas that filters the sun's ultraviolet (UV) radiation. A diminished ozone layer allows more radiation to reach the Earth's surface. For people, overexposure to UV rays can lead to skin cancer, cataracts, and weakened immune systems. Increased UV exposure can also lead to reduced crop yield and disruptions in the marine food chain, among other harmful effects.

---

2   A complete history of the development of CFC refrigerants is found in Seth Cagin and
    Philip Dray, *Between Earth and Sky: How CFCs Changed Our World and Endangered the Ozone
    Layer* (New York: Pantheon, 1993); Bernard A. Nagengast, "A Historical Look at CFC
    Refrigerants," *ASHRAE Journal*, November 1988: 37-39; Bernard A. Nagengast, "History of
    Sealed Refrigeration Systems," *ASHRAE Journal* 38.1 (1995): S37, S42-S46, S48; Thomas
    Midgley, Jr., "From the Periodic Table to Production," *Journal of Industrial Engineering
    Chemistry*, February 1937; and Thomas Midgley, Jr. and Albert L. Henne, "Organic
    Fluorides as Refrigerants," *Industrial and Engineering Chemistry* 22.5 (1930): 542-45.

into force; and CFC production was halted in developed countries and agreed to be halted in developing countries.[3]

When the Montreal Protocol was signed in 1987, Trane responded immediately. The engineers at Trane were dismayed by the discovery that the CFC refrigerants they used in their product were destroying the ozone layer, and they were determined to become one of the leaders in shifting to less harmful chemicals. Along the way they became advocates for global environmental protection, and for products that are the best in the world not only technically and economically, but also environmentally. In the late 1980s, Trane began its campaign to protect the ozone layer by eliminating the use of CFCs, the most harmful ozone-depleting refrigerants.

Think globally, act environmentally

Ozone-layer protection required fast action, but Trane managers knew that the ozone-safe HFC refrigerants proposed to replace CFCs were themselves potent greenhouse gases. Furthermore, the HFC refrigerants in commercial chillers are unable to achieve the highest energy efficiency, resulting in a further increase in greenhouse gas emissions from the power plants that supply the electricity to air-conditioning equipment. To solve this dilemma, Trane engineers embraced a broader accounting of environmental consequences to measure both the ozone-depleting potential and the global-warming potential of a refrigerant.

The comprehensive accounting by Trane clearly indicated that a shift to the new hydrochlorofluorocarbon (HCFC)-123 would be significantly less harmful to the environment, when damage to both the ozone layer and the climate were included.

HCFC-123 is the only commercially available, low-pressure refrigerant, and it has the highest energy-efficiency potential.[4] Its low pressure meant that HCFC-123 emissions could be very low because leakage through a hole of any size is directly proportional to pressure: low pressure, low leaks; high pressure, high leaks. Their theoretical energy-efficiency potential meant that HFC-123s could be harnessed to reduce their emissions from the power plants that generate electricity to run buildings' air-conditioning chillers.

In the meantime, scientific evidence was mounting that the 50% reduction called for under the 1987 Montreal Protocol was not enough to protect the ozone

---

3　See Stephen O. Andersen and K. Madhava Sarma, *Protecting the Ozone Layer: The United Nations History* (London: Earthscan Publications, 2002).

4　See Mark O. McLinden and David A. Didion, "Quest for Alternatives: A Molecular Approach Demonstrates Tradeoffs and Limitations Are Inevitable in Seeking Refrigerants," *ASHRAE Journal* 29.12 (1987): 32-36, 38, 40, 42; L.E. Manzer, "The CFC–Ozone Issue: Progress on the Development of Alternatives to CFCs," *DuPont Articles* (Wilmington, DE: DuPont) 249 (July 6, 1990): 31-35; James M. Calm and David A. Didion, "Trade-Offs in Refrigerant Selections: Past, Present, and Future," paper presented at *Refrigerants for the 21st Century*, ASHRAE/NIST *Refrigerants Conference*, Gaithersburg, MD, October 6–7, 1997.

layer. Technical assessments confirmed that air-conditioning equipment had very high refrigerant emissions from normal leaks, from frequent component failures and accidents, and from intentional venting during equipment servicing. The regulatory consensus was that the only sure way to avoid emissions was to phase out the production of ozone-depleting chemicals and to shift to new, environmentally acceptable chemical refrigerants.

In 1990, the Montreal Protocol was amended to phase out CFCs by 2000. In 1992, the Protocol was amended again to accelerate the CFC phase-out to 1996. The amended Protocol recognized HCFCs as important "transitional substances" necessary for the CFC phase-out but, nonetheless, it also scheduled an HCFC production phase-out by 2030 in developed countries, and by 2040 in developing countries.

In order to convince the leaders of the world to allow HCFC-123 to stay on the market, Trane and the rest of the heating, ventilating, and air-conditioning industry had to figure out how to use HCFCs in a sustainable and environmentally responsible way.

Trane engineers decided to redesign their chillers to prevent virtually any possible leak of HCFC-123. They also wanted to be able to document their near-zero emissions to help persuade environmental authorities, customers, and the public of the environmental soundness of their strategy. Trane's engineers started with a simple proposition: the best way to keep the refrigerant from escaping would be to eliminate as many openings as possible on the chillers where refrigerant could leak.

The squeaky wheel gets no grease

The oil system on Trane and most other chillers was an elaborate system of copper tubing, a heater to heat the oil, and a motor to circulate the oil to the right place in the chiller—the bearings on the drive shaft. Refrigeration chillers are either 'direct-drive' with the motor on the same shaft as the compressor or 'gear-drive' in order to turn the compressor faster or slower than the motor (see Figure 9.1).

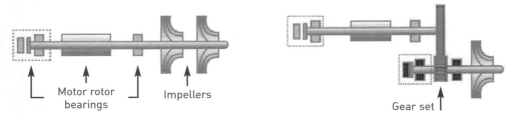

Direct-drive design   **Gear-drive design**

Figure 9.1 Comparison of direct- and gear-drive compressors

Although a great invention for its time, this complicated oil system was a potential source of refrigerant leaks. Trane engineers knew that eliminating the oil system would eliminate the major openings where refrigerant could leak out. They also knew that eliminating the oil would have the extra advantage of improving energy efficiency, because oil mixed with refrigerant degrades the chiller's performance. In fact, the advantages are so great that an oil-free chiller has long been considered the Holy Grail of the chiller business (see Figure 9.2).

But designing a system to run without oil had so many significant engineering hurdles that many of the most innovative engineers considered it impossible.

Or was it? Trane's vice president of technology, David Eber, believed that it was possible. King David (so named because he was considered "king" when it came to successful designs) had such a long and successful track record designing centrifugal chillers—playing an integral part in developing the succession of technologies that cumulatively increased their energy efficiency (see Figure 9.3)—that the company gave him wide latitude to decide what technologies to pursue. Eber knew that oil-free compressors had been tried over the years. He knew that magnetic bearings were one possible solution, but he also knew that those bearings cost up to $15,000—whereas steel bearings cost only $200. And other available solutions could not survive the start–stop–start requirements demanded by a chiller.

To solve the problem Eber thought back to 1963, when he was in his very first job with Caterpillar. He remembered working on a bearing problem in a gas

Traditional A/C chiller

Oil-free A/C chiller

Figure 9.2 Traditional and oil-free A/C chillers compared

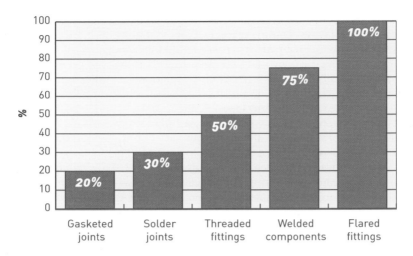

Figure 9.3 Evolution of refrigerant leak reductions

turbine engine and asking for help from one of the old hands, a German engineer who had previously worked with rocket scientist Werner von Braun. The German engineer told Eber that the solution was "carbide races and ceramic ball bearings." (See Figure 9.4. "Races" are the metal rings that the balls roll inside.) More than 30

Figure 9.4 Cut-away view of typical bearing balls and race

years later, Eber returned to ceramic ball bearings as a possible solution to Trane's problem. He was familiar with the ceramic balls used in machine-shop milling equipment, and he was optimistic.

But ceramic bearings were expensive, costing perhaps $2,000, which was ten times what the steel bearings cost. Eber knew that he could only use the expensive ceramic bearings if he could find some way to offset their higher cost. He knew that the ceramic bearings would need less oil than the old steel ones to run smoothly; but they would still need the same elaborate pump system to deliver the oil to the bearings. Eber thought about the oil system. He could save a lot of money if he could eliminate it altogether, and find some other way to provide the light lubrication the new bearings needed.

David Eber

With this goal in mind, Eber got his "bearing guys," Byron Hamm and Todd Smith, together and they started working with ceramic bearings in the lab with the help of their Swedish bearing supplier, SKF. They were able to get the ceramic bearings to work in the lab for the loads experienced in a building chiller. The ceramic balls are very light and incredibly hard—much harder than steel. Eber and his engineers like to show how hard the ceramic bearings are by giving visitors a hammer and a ceramic ball bearing, and telling them to put the bearing down on the cement sidewalk outside the headquarters and hit it as hard as they can. The ball bearing doesn't get a scratch, but the hammer is visibly dented and so is the sidewalk.

Ceramic bearings are hard; they don't pit or gall; and they have an extremely long life. Unlike steel bearings which can fail after being nicked or scratched, ceramic bearings also have the unique characteristic of "self-healing." Ceramic bearings buff away nicks and scratches in the bearing race.

Hamm and Smith thought about the challenge. Even with the ceramic bearings, they still saw two problems they would have to solve. The first was to reduce friction between the balls and the drive shaft without oil. The second was to remove the heat caused by whatever bearing friction was left.

Could they find another way to lubricate the bearings? One possibility was to use the one chemical that was already in the system, the refrigerant HCFC-123. But

> **NEW FROM TRANE**
>
> The new Trane S-Series EarthWise™ CenTraVac™ building air conditioner is the first high-efficiency, oil-free building air-conditioning chiller. The elimination of lubricating oil further increases the energy efficiency of Trane chillers, which were already the best in the world. Its oil-free, semi-hermetic design and advanced refrigerant recovery system make it possible to reduce refrigerant emissions almost to zero—eliminating the direct environmental impact of the refrigerant and making the chiller's high efficiency/near-zero-emissions operation sustainable throughout its entire lifetime.

what kind of lubricating potential did HCFC-123 have, and could Hamm and Smith devise a way to lubricate the bearings with it?

The Trane team was on a quest for better engineering. "I'm a gear head, like many of us here," said colleague Lee Cline. Others referred to themselves as "tinkerers," and you could see their eyes light up when they were in the machine shop at the R&D facility where they build and perfect their new chiller designs. These are people who like to get their hands on the machines. They run the most sophisticated computer models, but they believe even more in the reality of working with the machine they are building and testing.

The steel bearings in Trane's chillers are designed to last the lifetime of the chillers, which is 30 years or more. But they sometimes need servicing, and for the current chillers this means taking the machine apart. Some refrigerant can escape when the machines are taken apart for repairs, even when the service staff are careful to evacuate the refrigerant before they disassemble the chiller.

Hamm and Smith reasoned that the ceramic bearings might work with just the lubrication provided by the refrigerant in the chiller—*if* the refrigerant could be delivered directly to the bearings, *if* it had sufficient lubricating properties, and *if* it was able to carry away the heat from friction. That's a lot of ifs, so, even if they could figure out a way to accomplish all of that, they would have one more major hurdle. Trane is a business and that means it is not just about what is possible. Hamm and Smith also had to consider the problem of cost. Could they get the overall price of the chiller down low enough to be competitive in the market? Would removing the oil system save enough money to justify the higher-priced ceramic bearings? The Trane team would come back to the problem of cost time and again. As Smitty said, "*You have to make green to be green.*"

Hamm and Smith put the problem of cost aside for the moment, and in May of 1998 began to build an experimental chiller without oil. They calculated that the "lubricity" of HCFC-123 was adequate for the super-hard ceramic bearings, but they were still struggling to solve the problem of delivering the refrigerant directly to the bearings.

Their first effort to deliver the refrigerant to the bearings took the form of a dedicated pump that injected refrigerant between the ball bearings and the race. Eber gave an extra incentive to his colleagues by directing them to do the building scale test on the air-conditioning chiller in Building 12, the engineering lab. This meant that all of the engineers in the lab would know whether Hamm and Smith were successful. Failure would not only be costly and embarrassing, but it would also leave their co-workers to suffer in sweltering-hot working conditions.

On the day of the test in Building 12, Eber was confident but also prepared to carry on even with failure. "Engineering success," he is quick to say, "is not what you might expect, and engineers need to have a forgiving environment in which to operate. In many other organizations, if they learn something positive from an experiment, they call it success. But, if they learn something negative, they say it is failure." Eber continues: "But it's different at Trane. We know we're going to learn something in either case, and if our experiment doesn't work, we study the problem further to learn why. Maybe it's something we can fix easily. *All experiments are a success at Trane.*" But Eber also quickly points out that at Trane "what is not forgiven is if Trane is ready to go into production, and the engineers have made a mistake that they could have avoided."

King David may have been confident no matter what, but, as they started up the experimental chiller, Trane's lubrication expert, Art Butterworth, was still worried. If it failed, the building would heat up and he would have a lot of grumpy engineers to deal with. Hamm, Smith, and their colleague Art Butterworth made their last-minute check of the experimental chiller. They joked that, if the chiller froze up, someone had better come up with a few tons of ice to cool off the engineers.

Smith threw the switch, and the chiller started up. But they all knew it would start. The question was, would it continue to run? They listened closely for the sounds that would indicate failure. They monitored the machine's temperature to see if it was overheating. But the experimental oil-free chiller didn't make any strange noises, and it didn't run hot. It just ran. It made a beautiful sound, a sweet mechanical hum. "The darn thing was working and working well."

Hamm and Smith shut it down after a few

> **TRANE OIL-FREE EARTHWISE™ AIR CONDITIONERS**
>
> **"Creating the right atmosphere"**
>
> *Good for the consumer:*
>
> - Saves on electricity with improved efficiency
> - Eliminates performance degradation from oil build-up
> - Hermetically sealed to maintain optimal refrigerant charge
> - Longer life
> - Eliminates expensive hazardous waste and oil disposal fees
> - Increased reliability with less maintenance
>
> *Good for the planet:*
>
> - Almost-zero emissions of ozone-depleting chemicals
> - Energy-efficient for the reduction of greenhouse gases
> - Simple design makes recycling easy

hours and took it apart to see what the ceramic bearings looked like. They also read the computer output that measured the flow of the lubricant to the two sets of bearings. The ceramic bearings and the bearing race looked fine, but the collar that held the bearings in place on one of the three bearings was scarred, indicating inadequate flow of the refrigerant acting as the lubricant.

The team discovered that the injection of refrigerant between the bearings wasn't working because a "cross-wind" from the compressor blew the entire dose of refrigerant to just one bearing!

But, of course, the Trane way of engineering is built to thrive whether a test works or doesn't work. The team went back to the drawing board undisturbed. And they came up with a new design.

The second design was a success! Field trials would come next—on-site tests for the things that engineers simply cannot anticipate in the lab. And so the project continued.

,uading the Montreal Protocol
allow emission-free HCFC use

For the new Trane design to become a reality, though, a key piece of international diplomacy had to be accomplished. The use of HCFC-123 in the Trane design depended on the availability of that chemical. What was needed was a decision by parties to the Montreal Protocol to allow the manufacture of small amounts of HCFC-123 after 2030, so that customers would be encouraged to continue to select the most energy-efficient chillers and be able to depend on the availability of HCFC-123.

The Montreal Protocol allows the production and use of HCFC-123 because it replaces the more potent CFCs, allowing their faster phase-out. After the CFCs phase-out, though, HCFC-123 is next in line, with a production phase-out scheduled for 2030/2040. Chillers are very expensive, costing between $200,000 and $600,000, and have 30-year life-cycles. As a result, customers looking to buy one during the next few years will only want one that uses a refrigerant that they can be sure will be around for many years, just in case they need to recharge their chiller. The other leading CFC-replacement refrigerant, HFC-134a, is ozone-safe but is a potent greenhouse gas included in the basket of six gases to be controlled under a different international agreement, the Kyoto Protocol.

Diplomats and international policy-makers will have the task of carefully weighing all of the pros and cons of each technology to decide what will be allowed and what will be controlled. The technical experts who advise the Montreal Protocol are members of the Technology and Economic Assessment Panel, and the experts that advise the Framework Convention on Climate Change are members of the Intergovernmental Panel on Climate Change. In 1999, these two expert panels conducted a joint assessment and conceived "life-cycle climate performance" to measure the integrated direct and indirect greenhouse emissions of a product through all stages of production, use, and disposal.

> **❝ This move toward high-efficiency, low-emission systems is a win–win situation: a win for both business and for the environment. As we move into the third millennium, the emphasis on high-efficiency, low-emission products and systems will continue and will define the difference between winning and losing technologies. ❞**
>
> *Jerry Arndt, Trane Vice President and General Manager, upon accepting the prestigious 1998 EPA Climate Protection Award*

They also proposed "responsible use principles" to guide technology choice and minimize harmful emissions. Responsible use principles allow the use of technologies so long as the undesirable effects are minimized and the technology achieves higher environmental performance than its alternatives. Trane achieves responsible use with near-zero refrigerant emissions and very high energy efficiency, earning stakeholders' confidence that greenhouse gases are managed wisely.

The engineers at Trane believe that the superior environmental performance will persuade environmental authorities to endorse HCFC chillers. Trane believes that, by standing on these principles of responsible use and by relying on the scientific studies done by world-class atmospheric scientists, they can factually build a powerful case for the extension or elimination of the current HCFC-123 phase-out dates. The company believes that a choice *other* than to eliminate the current phase-out dates would actually bring more harm to the environment, rather than less harm. Figure 9.5 shows two of the journal articles that have asked for a reconsideration of HCFC-123 phase-out dates.

The HCFC-123 chiller emissions rates are already low. Trane engineers believe that they can be made even lower with an oil-free, hermetically sealed compressor, with microprocessor-based control and monitoring, and with state-of-the-art refrigerant recovery technology. Moreover, HCFC-123 chillers have significantly higher energy efficiency than equipment using the HFC refrigerants allowed by the Montreal Protocol (but controlled by the Kyoto Protocol).

Figure 9.5 Magazine articles urge Montreal Protocol to reconsider

Knowing that Montreal Protocol Parties will only act when they are confident of the environmental benefit, Trane Environmental Affairs Director Smithart persuaded Trane to implement every technically feasible measure to further reduce both refrigerant and utility-generated emissions as proof of Trane's commitment to the responsible use principles. Along the way, Trane has received numerous environmental achievement awards, including the prestigious EPA Climate Protection Award in 1998. The award acknowledged that the Trane EarthWise CenTraVac chiller has a proven record as the world's most efficient, lowest-emissions chiller.

Gene Smithart

The EPA citation stated:

> The Trane EarthWise CenTraVac centrifugal water chiller uses an environmentally balanced alternative refrigerant and leads the industry with superior performance, boasting efficiencies of 0.48 kilowatts per ton or better at ARI [Air-Conditioning and Refrigeration Institute]-rated conditions and a "near zero" refrigerant emissions level. Performance exceeds all other product technologies currently in the marketplace, typically by 5% to 20%.[5] *Further, it is 40% to 50% more efficient than chillers installed 15 to 20 years ago, creating a tremendous opportunity for both environmental and economic improvements* [authors' italics].

Making green

There has been a great deal of interest as news of Trane's achievement has spread around the world. Several major companies that are committed to the environment, and to their own pocket books, switched to Trane's new chiller. One of those companies is Connecticut Mutual Life Insurance Company, whose headquarters houses 1,200 employees and functions as the company's main data-support center for the entire USA—and which is committed to energy conservation. The company's priority was to select the chiller that would offer the greatest possible efficiency while assuring safe operation for its personnel. After examining the products in the marketplace, Connecticut Mutual decided on an EarthWise chiller from Trane.

5   The Sustainable S-Series EarthWise™ CenTraVac™ chiller utilizes ceramic ball bearings that are lubricated with refrigerant not oil. For more information, visit www.trane.com/commercial/equipment/earthwise_systems.

TRANE IMPROVEMENTS IN ODP, GWP, AND ENERGY EFFICIENCY

- Improved 1,000 ton chiller energy efficiency—typical savings of $1,440 per 1% improved efficiency
- Lower bearing frictional losses—up to 1% = $1,440 per year
- Eliminate performance degradation from oil build-up in refrigerant—up to 10% compared to older chiller designs = $14,400 per year
- Maintain optimal refrigerant charge with hermetic seal—up to 10% improvement
- Improved sustainable performance of oil-free surfaces in heat exchangers—potentially large
- Increased chiller reliability
- Longer bearing life
- Less frequent routine maintenance and expense
- Simpler equipment with fewer components
- Better containment avoids loss-of-refrigerant failures
- No oil management problems or hazardous waste oil disposal
- Enhanced by Trane with remarkable microprocessor-based technology and management tools
- An ultra-high-efficiency purge system recovers and automatically reclaims HCFC-123 when air is purged from the system.
- Automated electronic monitoring of 200 performance indicators helps keep the equipment operating at top efficiency, increases system reliability, and gives early warning of possible refrigerant leaks. This sophisticated monitoring also verifies and documents energy savings.

Washington Adventist Hospital also replaced three of its ageing air-conditioning chillers that run on CFC refrigerant with two 740 ton EarthWise CenTraVac chillers. The chiller replacement is part of the hospital's overall environmental policy, according to Kiltie Leach, the hospital's COO (chief operating officer). "The chiller replacement exemplifies our commitment to being an environmentally conscious and cost-efficient health care institution."[6]

The headquarters of the US Environmental Protection Agency also installed two EarthWise CenTraVac chillers, which were dedicated on Earth Day, 1995 (Figure 9.6). Because of their efficiency improvements, which include high-efficiency lights and new window treatments, the two 670 ton Trane chillers are far smaller than the two 875 ton machines they replaced. On a peak-load day, which can reach 95°F (35°C) in the summer, each chiller draws only about half the power that each older machine would have required.

Figure 9.6 EPA Ronald Reagan Building features Trane EarthWise chillers

6   "Washington Hospital Selects New Chillers, with New Refrigerant", *The Air Conditioning, Heating and Refrigeration News* (Business News Publishing Co.), April 3, 1995.

We *didn't say our chiller was*
environmentally
responsible.
The **EPA did.**

"The Trane EarthWise CenTraVac centrifugal water chiller is used to cool large office buildings, airports, and hospitals, as well as for industrial process applications. This chiller uses an environmentally balanced alternative refrigerant and leads the industry with superior performance, boasting efficiencies of 0.48 kilowatts/ton or better at ARI-rated conditions and a 'near zero' refrigerant-emissions level. Performance exceeds all other product technologies currently in the marketplace, typically by 5 to 20 percent. Furthermore, it is 40 to 50 percent more efficient than chillers installed 15 to 20 years ago, creating a tremendous opportunity for both environmental and economic improvements." — *U.S. Environmental Protection Agency Citation*

You see, the Trane® EarthWise™ CenTraVac™ chiller is proven to be the most-efficient, lowest-emission chiller on the planet. But don't just take our word for it.

Don't just take our word for it, take the EPA's...

The U.S. Environmental Protection Agency awarded the Trane EarthWise CenTraVac chiller its Climate Protection Award. It is the only chiller in the world to receive this prestigious award. Judged by an international panel of industry, government, and organizational leaders, the award recognizes Trane's leadership role in addressing climate-change issues across the globe.

The right refrigerant. The right results.

The key to its exceptional performance lies in the chosen refrigerant. Designed for use with HCFC-123, a low-pressure refrigerant, the EarthWise CenTraVac chiller attains peak energy-efficiency levels and lowest total emissions. HCFC-123 has the highest thermodynamic efficiency of

all non-CFC refrigerants and the lowest direct-effect global warming potential. And, it can be essentially recycled indefinitely. For these reasons, HCFC-123 is used in more new centrifugal chillers today than all other alternative refrigerants combined.

Lower emissions for your site and your energy supplier.

Besides having the industry's lowest refrigerant emission rate, the EarthWise CenTraVac chiller reduces emissions at the utility plant. By running at such high efficiency levels—as low as 0.45 kilowatts per ton—the EarthWise CenTraVac chiller helps reduce emissions of utility-generated greenhouse gases. Trane balances concerns pertaining to energy efficiency, ozone depletion, and global warming, all in one piece of equipment. Just think: if every chiller could operate at 0.45 kW/ton efficiency, utility greenhouse-gas emissions would be reduced by more than 21 billion pounds of CO_2 while reductions in SO_2 and NO_x would be more than 80 and 34 billion grams, respectively.

Interested?

If you want to do what's best for your building and the environment, you've come to the right place. Trane. For more information on the EarthWise CenTraVac, please visit **http://www.trane.com/commercial/equipment**.

TRANE

Figure 9.7 Trane advertisement featuring EPA climate award

TECHNICAL ACHIEVEMENTS

Large buildings in hot climates require large air conditioners called "chillers." Chillers are custom-built machines costing $200,000–600,000 that consume large amounts of energy over the lifetime of the building.

Only a direct-drive design can be oil-free, and only HCFC-123 can offer the highest-efficiency direct-drive designs.

Hybrid ceramic bearings can be oil-free, but gear sets require other lubrication. Because it has no gear sets, a direct-drive design—with impellers that are on the same shaft as the motor rotor—can be oil-free. The only available refrigerant suitable for the highest-efficiency direct-drive applications is HCFC-123, because it operates at significantly lower pressure than other commercially available refrigerants such as HFC-134a.*

Oil-free design exploits theoretical energy performance—avoids efficiency degradation.

Any oil mixed with refrigerant degrades energy performance, with greater loss of efficiency at higher oil concentration. For example, 8% oil costs 15% efficiency and 3.5% oil costs 8% efficiency. Even 1% oil costs 3% efficiency. Today's chillers are designed to minimize or recover oil from the refrigerant, but the best solution is to eliminate oil altogether.

Oil-free design has less friction and less heat.

Hybrid bearings increase compressor energy efficiency by up to 1%. Part of the gain is from reduced friction; another part of the gain is from needing less energy to cool oil.†

continued ➜

* The maximum theoretical coefficient of performance (COP) for HCFC-123 is 7% greater than for HFC-134a. This advantage comes from the shape of the saturation dome and the relative slope of constant entropy lines followed by the compressor.

$$\text{COP} = \frac{\dot{Q}_{evap}}{\dot{W}_{comp}}$$

$$\dot{Q}_{evap} = \dot{m}_R \cdot \left(h_{RV}^{sat}\left[T_{sat,E}\right] - h_{RL}^{sat}\left[T_{sat,C}\right] \right)$$

$$\dot{W}_{comp} = \dot{m}_R \cdot \left\{ h_{RV}\left(s_{RV}^{sat}\left[T_{sat,E}\right], P_c \right) - h_{RV}^{sat}\left(T_{sat,E}\right) \right\}$$

$$\text{COP} = \frac{\left(h_{RV}^{sat}\left[T_{sat,E}\right] - h_{RL}^{sat}\left[T_{sat,C}\right] \right)}{\left\{ h_{RV}\left(s_{RV}^{sat}\left[T_{sat,E}\right], P_c \right) - h_{RV}^{sat}\left(T_{sat,E}\right) \right\}}$$

† A simplified view of the total frictional moment in a ball bearing shows that it is comprised of two components, the load-independent moment M_0 and the load-dependent moment M_1,

$$M = M_0 + M_1$$

$$M_0 = 6.6 \times 10^{-7} \, (\nu n)^{2/3} \, d_m^3$$

$$M_1 = f_1 \, P_1 \, d_m$$

The total frictional improvement of a hybrid ceramic bearing as compared to an all-steel oil-lubricated bearing system can be represented as:

$$M_{hc} = 0.07 M_{0st} + 0.50 M_{1st},$$

$$F = K_f \mu \, N r \, l/m$$

Oil-free lubrication reduces maintenance cost.

Typical annual oil filter changes cost from $200 to $400 per year—$6,000 to $12,000 over 30 years of a chiller's operation. An oil change for HFC chillers using POE (polyol ester) oil costs an additional $600–1,000; $1,200–6,000 for two to six typical oil changes over the lifetime of the chiller. An oil change for HCFC-123 chillers that use mineral-based oils typically costs an additional $400–800 per oil change—$800–4,800 for two to six typical oil changes over the lifetime of the chiller. Chillers that use oil-based lubrication have an electric oil heater to extract refrigerant from the oil and to assure proper viscosity during cold-temperature operation. A typical 750 watt heater operating 4,000 hours at $0.08 per kWh costs $240 per year. Additional energy is used to cool the hot oil after refrigerant is evaporated.

The initial refrigerant charge lasts a lifetime and will be recovered and re-used.

The Trane S-Series chiller is hermetically sealed, and has far fewer joints, gaskets, and fittings than other machines. With no oil filters or oil to change, the refrigerant loss during routine maintenance is eliminated. In addition, sealed systems will always operate at optimal refrigerant charge for highest energy efficiency and lowest emissions of utility-generated greenhouse and toxic combustion gases. Systems that leak can lose 25–45% of their refrigerant charge before shutting down—resulting in up to 15% more energy use for prolonged periods of time.

HCFC-123 has higher theoretical and achieved efficiency.

HCFC-123 for air-conditioning applications is 5–20% more efficient than is HFC refrigerant used in air-conditioning chillers. The highest efficiency of comparable HCFC-22 and HFC-134a chillers is approximately 0.54 kW per ton, while the entire EarthWise chiller product line achieves at least 0.49 kW/ton with some sizes as low as 0.45 kW/ton.[‡]

High energy efficiency helps protect the climate and local air quality.

If every centrifugal chiller in the world operated at 0.45 kW per ton efficiency—a rate only currently obtainable with HCFC-123—annual power plant emissions would be reduced by over 17.7 billion lb (8 billion kg) of CO_2, over 160 million lb (72 million kg) of SO_2, and over 66.7 million lb (30 million kg) of NO_x. *This is equal to removing over 2 million cars from the road, or planting more than 500 million trees.*

Monitored and documented performance

The S-Series EarthWise™ chiller monitors and documents the key information needed to ensure high financial and environmental performance, automatically giving operators early warning of any problems. For example, increasing purge run time is an indication of minute refrigerant leaks. Changing condenser and evaporator approach temperatures or power consumption indicates possible tube fouling or water distribution problems. Evaporator and condenser temperature differential indicates capacity and accuracy of control. All of these are monitored.

The chiller is part of a system.

Total building energy efficiency is achieved by thorough attention to all building energy loads and building systems, including the heating and cooling systems. Chapter 4 on Energy Star describes how the purchase of energy-efficient office equipment and lighting can reduce the building heat load, reducing the energy consumption of air conditioning.

‡ Because the first commercial refrigeration was with ice, the unit of measure for cold is a ton of ice. A kilowatt-hour is the standard measure of electricity and is one kilowatt of power supplied continuously for one hour.

EPA's chiller installations, along with its other energy-saving steps, will mean savings of more than 4 million kWh of electricity per year, according to Trane. At typical electricity costs of around $0.075 per kWh, savings will be approximately $300,000 per year, it said.[7]

"With EPA saving 5 million kWh of electricity a year, utility-generated emissions will be reduced each year by 6 million pounds [2.7 million kg] of carbon dioxide, 72,000 pounds [32,400 kg] of sulfur dioxide, and 22,000 pounds [9,900 kg] of nitrogen oxide," explained Jim Schultz, then executive vice president of Trane's North American Commercial Group.[8] But it's not just about being green. The day Trane accepted the EPA award, it was the market leader, with worldwide installations numbering in the tens of thousands. It's about making green, for the company and for its customers.

Trane time-line

1938 ● World's first direct-drive, hermetic, multi-stage, centrifugal water chiller.
1981 ● First three-stage, direct-drive, centrifugal chiller.
1992 ● EPA Stratospheric Ozone Protection Award.
1993 ● EarthWise CenTraVac, the world's most efficient lowest-emissions chiller.
1995 ● Wisconsin Society of Professional Engineers Award for New Products—
       EarthWise CenTraVac.
       ● EPA Stratospheric Ozone Protection Award for James Wolf.
1998 ● EPA Climate Protection Award for the EarthWise CenTraVac.
       ● EPA Energy Star Buildings Ally of the Year.
1999 ● EPA Climate Protection Award for Eugene Smithart.
2002 ● S-Series, Oil-Free, EarthWise CenTraVac.

---

7 For example: a 1,000 ton chiller rated at 0.576 kW/ton (American Society of Heating, Refrigeration and Air-Conditioning Engineers [ASHRAE] 90.1 minimum efficiency) with 2,000 equivalent full-load operating hours (EFLH) during a six-month cooling season, paying $0.08/kWh and a $15/kW demand charge, has total operating costs of $144,000 per year. Each 1% of operating costs is $1,440 per year (1,000 tons × 2,000 EFLH × 0.576 kW/ton × $0.08/kWh = $92,160; 1,000 tons × 0.576 kW/ton × $15/kW × 6 months = $51,840 demand charge).
8 Interviewed by Stephen Andersen.

10
Visteon
Superintegration™
proves "simple is better"*

A fresh design paradigm integrates flexible electronic circuits to reduce materials, space, and weight, and at the same time improve the reliability and energy efficiency of automobiles.

Visteon's new Superintegration™ is a design strategy that minimizes the number of electrical components and connections in automobiles. Combined with proprietary technology, it allows electronic circuits to be integrated on structural and functional surfaces—such as the underside of a dashboard. It also allows circuits to be incorporated within flexible cable bundles that include electronic components and replace old, single-function printed circuit boards and old-fashioned wire harnesses.

Superintegration™ has many benefits. By combining no-clean and lead-free soldering with flexible circuits integrated directly into structural surfaces, it:

- Reduces weight of vehicles including cars, trucks, and airplanes
- Reduces materials use
- Saves operating energy
- Simplifies assembly and disassembly
- Accelerates electronic functions

Cars produced with this technology have dramatically improved reliability. In addition, the technology offers significant weight reduction which improves fuel economy.

  *   The authors are grateful for interviews and supplementary assistance by Visteon engineers and managers Jay Baker, Larry Kneisel, Myron Lemecha, Sheila Lowe-Burke, Rick McMillan, Brenda Nation, and Charles Schweitzer.

Superintegration™ also provides design and manufacturing flexibility which allows solutions to logistical problems that were previously impossible to solve. For example, Superintegration™ allows cockpit management electronics to be placed directly on the cross-car beam structure, speeding communication between computer controls and vehicle functions, and simplifying fault diagnoses and repair. Faster electronic circuits, in turn, allow more sophisticated management of cockpit systems, which also results in significant weight reductions.

With this technology, the cockpit unit becomes fully self-contained and modular, allowing a manufacturer to upgrade more quickly as technology improves. Production applications have demonstrated savings of up to 30% of the system weight and as much as a 50% part-count reduction (see Table 10.1 on page 181). As a result, Superintegration™ will help protect the climate when used for automotive, electronics, aerospace, and other applications by reducing the mass of the vehicle that has to be propelled and therefore the overall energy used for propulsion.

This design strategy also eliminates all of the bells and buzzers that typically alert automobile drivers to conditions such as "door open," "unfastened seatbelt," or "lights on" and replaces them with sounds generated on the car audio system speakers. As a result, less energy is embedded in materials, components, and assembly. The simplified assembly and disassembly that this enables also encourages the recovery and re-use of components and materials, contributing further to the protection of the climate.

The story of Superintegration™

Superintegration™ was born of innovations designed to protect the ozone layer. Prior to the signing, in 1987, of the Montreal Protocol for the protection of the ozone layer, the ozone-depleting solvent CFC-113 was used widely in electronics manufacturing for the post-solder defluxing of printed circuit boards. This CFC solvent was an obvious choice because of its excellent cleaning properties, low toxicity, non-flammability, and relatively low cost.

But, after the Montreal Protocol, companies in the electronics industry were galvanized to investigate alternative cleaning methods. These companies faced a challenge: they needed a way to clean resin off electrical components completely, because otherwise, after it was used to help solder stick and to de-oxidize components, it became a corrosive and conductive residue that would ultimately destroy the electronics.

In 1989, Jay Baker was assigned to a new position at Ford Motor Company. His mandate was to develop advanced manufacturing processes for electronics systems; and he had the added responsibility of diagnosing and eliminating electronic failures that were costing large amounts of money in warranty repair.

> **DEFINITION OF SUPERINTEGRATION™**
>
> Superintegration™ is a group of breakthrough technologies that enable physical and functional integration of electronics in systems. It uses the optimal combination of materials for products that lead the way to greatly improved function, performance, manufacturing, and cost.

EFFECTIVENESS AND ENVIRONMENTAL ADVANTAGES OF SUPERINTEGRATION™

The "Plug-and-Play" feature will allow upgrades to navigation systems that save fuel by using real-time information to route around congested traffic, and they could allow owners to take advantage of the latest computer ignition management systems that minimize combustion emissions.

Furthermore, the increased modularity of Superintegration™ will make it more technically and economically feasible to re-use components and to recycle materials. This is enabled because entire assemblies will be separable, because the assemblies will be fabricated from a smaller number of plastics (reducing identification and sorting), and because removable parts will be interchangeable between more vehicles.

Efforts to reduce pollution and emissions of greenhouse gases can be strong incentives for the adoption of Superintegration™ technologies. The voluntary agreement of vehicle manufacturers in Europe to increase fuel efficiency by 25% by 2008 will encourage technological reductions in vehicle weight. The European requirements for end-of-life product stewardship and mandatory recycling rates will encourage modularity and reductions in the types of plastics used.

Ford's corporate office asked Jay to attend a meeting organized by EPA to discuss cooperation in commercializing alternatives to CFC solvents; but he did not receive much other corporate support for that undertaking. "You're on your own," said the corporate office. After several meetings, the EPA-sponsored group decided to form an organization that would later be named the Industry Cooperative for Ozone Layer Protection (ICOLP).[1]

Baker stepped up as an early leader in the group, despite an apparent lack of support from other car companies. "The GM man was on the fence, leaning toward declining, and the Chrysler man said Chrysler was not going to join. IBM was also skeptical about joining. AT&T was for it. DEC was for it. Motorola was for doing it. I did a quick survey at the meeting. It was about 50–50," says Baker.[2] Ford was the only vehicle manufacturer to join.[3]

Baker recalls his thoughts at the time:

> Innovation is greatest when companies have engineers as top managers. Ford's Director of Technology, Charles Szuluk, was our god. He knew the bottom line, but he also knew who to trust and how to get things done. As a smoker, he would often join line workers in designated smoking zones outside the building where he could share a smoke and listen to their candid comments. He always came back with a list of things that needed fixing.

---

1   See Center for International Environmental Law (CIEL), *The Industry Cooperative for Ozone Layer Protection: A New Spirit at Work* (Washington, DC: CIEL, 1994).

2   Jay Baker, quoted in Penelope Canan and Nancy Reichman, *Ozone Connections: Expert Networks in Global Environmental Governance* (Sheffield, UK: Greenleaf Publishing, 2002): 20.

3   Founding members included AT&T, The Boeing Company, Digital Equipment Corporation (DEC), Ford, General Electric, Honeywell, Motorola, Nortel, and Texas Instruments. British Aerospace, Compaq, Hitachi, Hughes Aircraft Company, IBM, Lockheed-Martin, Mitsubishi Electric, Ontario Hydro, and Toshiba joined later.

Figure 10.1 Typical direct injection without Superintegration™

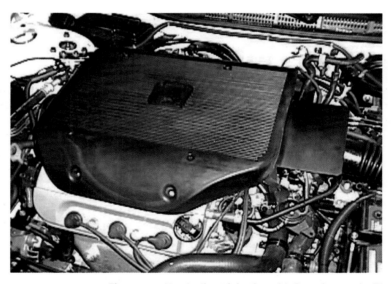

Figure 10.2 Honda direct injection with Superintegration™

ICOLP companies formed teams to follow this strategy and investigate each possible alternative to CFC solvents.

This was the time when the key difference emerged in Baker's thinking. Everyone was asking how to clean in a new way. But Baker asked a different question: "Why clean?" And he formed a team with AT&T, IBM, Motorola, Nortel, and Texas Instruments to investigate.

Some of Jay's team concentrated on the flux composition; others worked to apply the flux to the board more precisely; and still others concentrated on the mixture of gases in the soldering chamber of a new process called inert gas soldering. The team consulted with flux suppliers who grasped the opportunity for developing and commercializing new products and intensified development in cooperation with the electronics manufacturers. The EPA encouraged the work by documenting and publishing the advantages for the global environment of having "no-clean" technology.

Meanwhile, engineers at AT&T Bell Laboratories were developing state-of-the-art spray fluxing machines to apply the optimal amount of flux precisely to the targeted locations on the printed circuit boards. Nortel was developing equipment to verify flux concentrations on production boards; and Motorola was experimenting with soldering ultra-miniaturized circuits with hybrid components, including optical devices and flexible connectors. One by one, companies satisfied internal quality controls and moved from lab scale, to pilot, and finally to full implementation. During implementation, experts from the inter-company team continued to cooperate to debug operations and optimize performance.

It is impossible to say just when the engineering team realized that their no-clean technology would revolutionize electronics assembly by exceeding the quality, reliability, and speed rates possible with conventional soldering. Engineers who had cautiously reported "as-good" performance began to report *improved* performance. Line managers cautiously increased the speed of soldering to rates never achieved with conventional soldering and found that in some cases defect rates actually *decreased.*[4] Ford ended up with a soldering process that is virtually defect-free, saving hundreds of millions of dollars.

ALL THE WAY IN SUPERINTEGRATION™

The evolution leading to Superintegration™ can be described as follows:

Level 1 Module assemblies—individual components, interactive systems. (5–10% improvement in cost and quality)

Level 2 Partial component integration—integrated mechanical designs, integrated electronics (10–20% improvement in cost and quality)

Level 3 Superintegration™ integration across physical/functional boundaries

Level 4 Superintegration™ with maximum application of semiconductor and integrated circuits, assembled on flatwire cables, totally eliminating unnecessary materials and enabling "Plug-and-Play" upgrades to all systems

---

4   Stephen O. Andersen, Clayton Frech, and E. Thomas Morehouse, *Champions of the World: Stratospheric Ozone Protection Awards* (EPA430-R-97-023; Washington, DC: US EPA, August 1997).

	Concept development SI cockpit		Customer applications SI cockpit			
	1998 ranger	1999 expedition	2001(a)	2001(b)	IPCS manifold	Package tray
Price reduction	10–20%	5–15%	17%*	14%*	5–15%	5–15%
Weight reduction	17%	15%	17%†	11%†	11%	31%
Parts count reduction	44%	30–50%	28%	28%	25%	44%
Quality/reliability	23–28%	22%	22%	–	20–40%	15–20%
Serviceability	28%	25–50%	35–50%	–	10%	10%
Installation labor	15–25%	15–25%	15–25%	–	20%	15–25%
NVH‡ improvement	10%	10%	10%	–	5–15%	3%

* 17% electronic materials
† And freed enough package space to incorporate all rear electronics into cockpit
‡ Noise vibration harshness

Table 10.1 Cost and quality improvements from Superintegration™

Baker's team

With CFCs eliminated, most companies celebrated their success and disbanded their teams. But Jay Baker couldn't leave well alone. "Our experience eliminating CFCs taught us to ignore conventional wisdom and to look for entirely new ways to simplify and improve. And our success in protecting the ozone layer created a desire to solve other daunting global problems," he said. So Baker kept his team.

Baker is proud of the personnel that he has brought to his team and says:

> We strive for diversity as part of equal rights but we also recognize the business case. People from different social and cultural backgrounds think differently, and smart people can be found in all disciplines. Our small team has experts with advanced degrees in computational fluid dynamics, chemistry, electrical engineering, and literature—anything that sparks the imagination.

He adds with a laugh, "No one understood why we needed someone in fluid dynamics." But Jay knew they did, and with persistence he got his way.

The team recruitment strategy is to find people who have already accomplished the impossible—with faith, persistence, and confidence. Team members will tell you that they are encouraged to stretch the technology beyond previous limits, to set goals that require drastic—not incremental—improvement.

Jay Baker's team. *Standing* (left to right): Myron Lemecha,
Brenda Nation, Jay Baker, Rick McMillan;
seated (left to right): Sheila Lowe-Burke, Larry Kneisel

Team meetings have a feel of confidence and openness. Larry Kneisel will tell you that the team can make anything work—but that every member is necessary for ultimate success. Rick McMillan is confident that customers will profit no matter what part of Superintegration™ they embrace. Myron Lemecha is proud that he earns money for Visteon "every day" with no-clean soldering, reliability, and manufacturing innovation. Brenda Nation credits Charlie Szuluk for being willing to take risks in developing breakthrough technologies. Chuck Schweitzer reports that he has no time to worry about the risk of failure because intense time pressure keeps everyone focused. Long-time team member Sheila Lowe-Burke, says: "These engineers are brilliant, good-hearted, never say 'can't', have total faith in what they are doing, and give it their best every day."

The Baker team has benefited from Ford's strong commitment to funding research and development. Ford has an "innovation fund" to move products forward to the point where they become profitable enough to support further research and development.

THE VISTEON TEAM IS CLEAR ABOUT WHAT IS NECESSARY FOR REAL BREAKTHROUGHS:

- Genius teams with confidence, funding, and motivation
- Top management with vision and daring
- Customers in pursuit of excellence, knowing what they want, and trusting in success
- Suppliers and partners complementing the core team
- Promoters taking the technology from proof to sales

The story of Visteon

Visteon was launched in 1997, and became independent from Ford in 2000. Visteon rewarded Jay with a license to proceed. "Ford wanted breakthrough technology. There is a continuous pursuit of reliability; technical innovation sells cars; and technical reputation and sales success attracts investors," remembers Baker.

The next breakthrough occurred when Jay's group looked at the electronics of an entire car—rather than looking at individual systems within the car, as had always been done before. They began by placing every electronic component on a large yellow table the size of a car. The number of components, the miles of interconnecting wire, and the hundreds of connectors astounded everyone. "For several days we just sat back and reflected on what we had discovered," he says. The problems were obvious:

- Too many connectors. Connectors flex with temperature change and as vehicles travel on rough and uneven surfaces. Flexing creates intermittent or permanent faults in electrical systems, sometimes jeopardizing safety. Jay calls connectors the Achilles' heel of electronics. "What's the first thing you do when an electronic appliance isn't working? You jiggle the connectors."

- Too much wire. Copper wire is expensive and heavy. Wire "harnesses" snake their way through holes drilled in body parts, weakening the structure and inviting air and water leaks.

Figure 10.3 The yellow table containing the electronics components of an entire car

- Too many small parts. It is difficult to find mounting locations for small parts, and each attachment invites future rattles and squeaks. Further, each mounting adds weight to the car and makes access during repair more difficult.

- Too much duplication. Every electric system—ignition, lighting, brakes, air bags, locks, windows, and entertainment—required new transformers, buzzers, fuses, and more.

- Too much interference. Every system needed additional shields to protect against the electromagnetic interference of other systems. Radios were shielded against ignition noise, airbags were shielded against speaker noise, and anti-lock brakes were shielded against voltage fluctuation caused by electric seats.

Baker wanted solutions to these problems, and he was willing to adopt innovative strategies to get them. "From the start we saw this as a merger of financial and environmental strategy. We brainstormed solutions. I told my engineers to ignore technical limits and to imagine the perfect assembly—to visualize the goal."

Ultimately they decided to simplify, combine, and re-engineer (Figure 10.4). They simplified by placing system components closer together, ideally within one semiconductor. They combined by having one component do the work of many (such as by sending all sounds through the car speakers, or by using shared transformers). They re-engineered by eliminating mounting boards and learning how to incorporate components directly into wiring, and by learning how to mount them directly on body surfaces.

Now they had to apply these innovations to real products. "We had the ideas, but we still needed project financing, and we needed customers to buy in," Baker remembers. Here's how he tells the story:

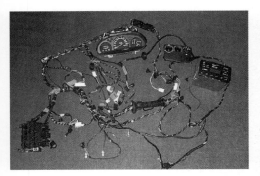

Conventional

Superintegrated

Figure 10.4 Superintegration™ cleans and simplifies electronic assembly

First we went bottom up, going door to door to each of the systems offices. The ignition people refused to share power-supply regulators, arguing that it would slow their work. The anti-lock brake people couldn't trust the entertainment people to keep their speakers (electromagnetic interference) from their sensors. The air bag people agreed that parts were redundant but felt it wasn't worth the trouble. Worse still, if management couldn't "see that it was broken," they didn't need to fix it. We were striking out.

Next we went cost-wise, going to purchasing agents seeking clout. But no one office at vehicle companies had responsibility for more than a few systems. They viewed joint procurement as an impossibility.

Finally we went top down, inviting executives to view our "wiring harness" hanger [see Figure 10.5] and our "yellow board" [Figure 10.3]. Executives entering the meeting saw a massive wire harness hanging from an overhead scale. They couldn't believe that it was all from one vehicle and they simply couldn't imagine that it could weigh that much. And when they looked at the yellow board their jaws dropped open. Of course, they agreed with the diagnoses and urged us to seek the cure.

At that point, Baker's team took advantage of its collaboration with ICOLP and the no-clean soldering project to organize a study tour of innovations in electronics assembly. The idea was that, once companies worked together successfully on environmental issues, they would become more open and more innovative. So the study tour visited centers of manufacturing excellence in Europe, North America, Japan, Russia, and Scandinavia.

Europe and North America have the advantage of having highly financed and motivated

> The wiring harness is an ugly collection of all of the wiring and electrical components from an automobile, hung from a harness so it can be weighed easily (Figure 10.5). The yellow board looks like a table top, and is about the size of the floor of an automobile; it has all of the wiring and electronic parts laid out as though they were in a car (Figure 10.3).

academic and government laboratories. Scandinavia has engineers with entirely different training, often motivated to go beyond product requirements and think about social and environmental performance. The Japanese approach to engineering is more focused on big-picture full integration. Their philosophy is to look beyond your assignments and to watch for ways to improve all the vehicle technology. Russian engineers are better trained in physics and mathematics and are more resourceful in solving problems with small budgets and limited access to computer and test facilities.

Baker says that their study tour was ten times more cost-effective and stimulating than anything they could have done themselves. For example, Seiko Epson hosted them in Suwa and shared their micro-assembly techniques, including their approach to connecting the mechanical and electrical components in their most sophisticated AGS (Automatic Generating System) watch and display components. And the tour went beyond looking just at technology; it also explored government–industry relations and engineering approaches. The tour was an occasion to see and learn and to be in the company of brilliant engineers—and it was also an opportunity for the Baker team's best engineers to spend quality time focused on

Figure 10.5 Vehicle wiring harness hanging on scale

problem-solving. "No phone calls, no conflicting meetings. Just a stream of consciousness, questions, and imagineering," as they described it.

Baker and his Visteon team turned "imagineering" into a process. Confident that no one else had accomplished their vision, they began fabricating and prototyping. When equipment suppliers told them that what they wanted couldn't be done, the team members rolled up their sleeves and designed their own equipment. When engineers at supply companies couldn't get the equipment to function properly, the Baker team brought in autoworkers from their own plants. So far they have earned *over 185 patents*, with dozens more pending. Superintegration™ grew from a reliability and environmental goal to an engineering approach and finally to an entirely new technology (Figure 10.6).

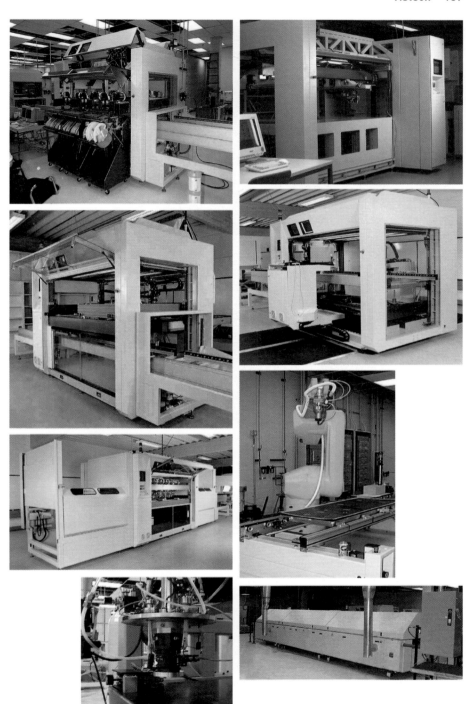

Figure 10.6 Visteon-invented Superintegration™ production machines

Current

Superintegrated cockpit

Figure 10.7 Before and after Superintegration™

Says Baker:

> Everyone who visits our development laboratory goes away with new skills. Visualize your own product on a "yellow board." Astound management by demonstrating how large and ridiculous systems have grown. Integrate and think outside the box. Of course, if our sales pitch is successful they also go away with an appreciation for what we have accomplished and, hopefully, with the decision to hire us to take them all the way in Superintegration™.

Visteon time-line

1799 ● Alessandro Volta (France) develops the first electrical battery.

1859 ● Gaston Planté (France) invents the first lead acid rechargeable battery, enabling automotive electronics.

1865 ● Alexander Parkes (England) discovers Parkesine, the first synthetic plastic.

1913 ● Henry Ford (USA) introduces the conveyer-belt assembly line to produce automobiles.

1987 ● Montreal Protocol is signed.

1989 ● ICOLP founded.

1990 ● Jay Baker is one of 16 individuals earning the first EPA Stratospheric Ozone Protection Award.

1992 ● A team of Brazilian soldering experts at Ford Brazil implements one of the first no-clean soldering technologies and transfers this advanced technology to developed and developing countries.

1994 ● Center for International Environmental Law (CIEL) publishes *The Industry Cooperative for Ozone Layer Protection: A New Spirit at Work*, praising the aerospace and electronics industry for global collaboration to eliminate ODS solvents.

1997 ● Visteon launched.

● Jay Baker earns the prestigious EPA Best of the Best Stratospheric Ozone Protection Award presented to those previous award winners who made the most exceptional contribution in the first decade of the Montreal Protocol.

2000 ● Visteon becomes independent from Ford.

Abbreviations

A/C	air conditioning
AC	alternating current
ACEEE	American Council for an Energy Efficient Economy
AGS	Automatic Generating System
AIST	Advanced Industrial Science and Technology (Japan)
ANPR	Advance Notice of Proposed Rulemaking
APB	Aviation Partners Boeing
API	Aviation Partners Incorporated
ARI	Air-Conditioning and Refrigeration Institute
ASHRAE	American Society of Heating, Refrigeration and Air-Conditioning Engineers
BBJ	Boeing Business Jets
BP	before the present
C_2F_6	hexafluoroethene
CART	Championship Auto Racing Teams
CD	compact disk
CD	critical dimension
CEO	chief executive officer
CF_4	carbon tetrafluoride (tetrafluoromethane)
CFC	chlorofluorocarbon
CH_4	methane
CIEL	Center for International Environmental Law
CO	carbon monoxide
CO_2	carbon dioxide
COO	chief operating officer
CPU	central processing unit
CRP	Cooperative Research Program (of SAE)
CVCC	Compound Vortex Controlled Combustion
CVT	continuously variable transmission
CWPB	center-worked, pre-bake
DDT	dichlorodiphenyltrichloroethane
DEC	Digital Equipment Corporation
DOE	Department of Energy (USA)
DTIE	Division of Technology, Industry and Economics (UNEP)
DVD	digital versatile disk
EFLH	equivalent full-load operating hours
EIA	Electronic Industries Association
EMAS	Eco-management and Audit Scheme
EPA	Environmental Protection Agency (USA)
EPROM	erasable programmable read-only memory
ESH	Environmental Safety and Health
ESIA	European Semiconductor Industry Association
FAA	Federal Aviation Administration (USA)
FBI	Federal Bureau of Investigation (USA)
FBO	Fixed Base Operation
FCHV	fuel cell hybrid vehicle

FI	fuel-injected
GHG	greenhouse gas
GM	General Motors
GWh	gigawatt-hour
GWP	global-warming potential
HC	hydrocarbon
HCFC	hydrochlorofluorocarbon
HFC	hydrofluorocarbon
HFE	hydrofluoroether
HSS	horizontal-stud Soderberg
IAI	International Aluminium Institute
ICOLP	Industry Cooperative for Ozone Layer Protection
ILEV	Inherently Low-Emission Vehicle
IMA	integrated motor assist
IP	Internet Protocol
IPCC	Intergovernmental Panel on Climate Change
ISO	International Organization for Standardization
JICOP	Japan Industrial Conference for Ozone Layer Protection
kWh	kilowatt-hour
LCCP	life-cycle climate performance
LCD	liquid-crystal display
LEV	Low-Emission Vehicle
MAC	mobile air conditioner
MACS	Mobile Air Conditioning Society
MBA	Master's in Business Administration
METI	Ministry of Economy, Trade, and Industry (Japan)
MIT	Massachusetts Institute of Technology
MITI	Ministry of International Trade and Industry (Japan)
MMt	million metric tons
MP3	Moving Picture Experts Group Layer-3 Audio
MPEG	Moving Picture Experts Group
mph	miles per hour
MT	million tonnes
N_2O	nitrous oxide
NAS	National Academy of Science
NASA	National Aeronautics and Space Administration (USA)
NEDO	New Energy and Industrial Technology Development Organization (Japan)
NGO	non-governmental organization
NGV	Natural Gas Vehicle
NH_3	ammonia
NIMC	National Institute of Materials and Chemical Research (Japan)
NLEV	National Low-Emission Vehicle
NMHC	non-methane hydrocarbon
NMOG	non-methane organic gas
NO_x	nitrogen oxides
NREL	National Renewable Energy Laboratory (USA)
NVH	noise vibration harshness
ODP	ozone-depletion potential
ODS	ozone-depleting substance
PFC	perfluorocarbon
POE	polyol ester
ppt	part per trillion
R&D	research and development
RACE	Refrigeration and Automotive Climate under Environmental Aspects
RITE	Research Institute of Innovative Technology for the Earth (Japan)
rpm	revolutions per minute

SAE	Society of Automotive Engineers
SF_6	sulfur hexafluoride
SO_2	sulfur dioxide
SoC	system on chip
SO_x	sulfur oxides
SPL	spent pot lining
STC	Supplemental Type Certificate
SULEV	Super Ultra-Low-Emission Vehicle
SUV	sport utility vehicle
SWPB	side-worked, pre-bake
TEAP	Technology and Economic Assessment Panel (UNEP)
TES	Technical Institute of Engineering Study (Germany)
TEWI	total equivalent warming impact
THC	total hydrocarbons
TT	tourist trophy
TV	television
TWh	terawatt-hour
UCLA	University of California at Los Angeles
ULEV	Ultra-Low-Emission Vehicle
UN	United Nations
UNEP	United Nations Environment Program
UV	ultraviolet
VAIP	Voluntary Aluminum Industry Partnership (EPA)
VCR	videocassette recorder
VP	vice president
VSS	vertical-stud Soderberg
VTEC	variable valve timing and lift electronic control
Vth	threshold voltage
WSC	World Semiconductor Council
ZLEV	Zero-Level Emission Vehicle
ZrO_2	zirconia